T0239066

SpringerBriefs in Earth System Sciences

Series Editors

Gerrit Lohmann, Universität Bremen, Bremen, Germany

Lawrence A. Mysak, Department of Atmospheric and Oceanic Science, McGill University, Montreal, QC, Canada

Justus Notholt, Institute of Environmental Physics, University of Bremen, Bremen, Germany

Jorge Rabassa, Labaratorio de Geomorfología y Cuaternar, CADIC-CONICET, Ushuaia, Tierra del Fuego, Argentina

Vikram Unnithan, Department of Earth and Space Sciences, Jacobs University Bremen, Bremen, Germany

SpringerBriefs in Earth System Sciences present concise summaries of cutting-edge research and practical applications. The series focuses on interdisciplinary research linking the lithosphere, atmosphere, biosphere, cryosphere, and hydrosphere building the system earth. It publishes peer-reviewed monographs under the editorial supervision of an international advisory board with the aim to publish 8 to 12 weeks after acceptance. Featuring compact volumes of 50 to 125 pages (approx. 20,000—70,000 words), the series covers a range of content from professional to academic such as:

- A timely reports of state-of-the art analytical techniques
- bridges between new research results
- snapshots of hot and/or emerging topics
- literature reviews
- in-depth case studies

Briefs are published as part of Springer's eBook collection, with millions of users worldwide. In addition, Briefs are available for individual print and electronic purchase. Briefs are characterized by fast, global electronic dissemination, standard publishing contracts, easy-to-use manuscript preparation and formatting guidelines, and expedited production schedules.

Both solicited and unsolicited manuscripts are considered for publication in this series.

The Research Group on Development Strategy of
Earth Science in China

Past, Present and Future
of a Habitable Earth

The Development Strategy of Earth Science
2021 to 2030

The Research Group on Development
Strategy of Earth Science in China
Beijing, China

ISSN 2191-589X   ISSN 2191-5903 (electronic)
SpringerBriefs in Earth System Sciences
ISBN 978-981-19-2782-9   ISBN 978-981-19-2783-6 (eBook)
https://doi.org/10.1007/978-981-19-2783-6

Jointly published with Science Press
The print edition is not for sale in China mainland. Customers from China mainland please order the print
book from: Science Press.

This Springer imprint is published by the registered company Springer Nature Singapore Pte Ltd.
The registered company address is: 152 Beach Road, #21-01/04 Gateway East, Singapore 189721,
Singapore

Editorial

Preface

The curiosity and spirit of exploration in understanding the origins of the Earth and life is the foundation of modern science, and this has never changed from ancient Greek civilization, the Renaissance, to the Industrial Revolution. Looking back at the development of geosciences in the twentieth century, we realized that it was the emergence of the theory of plate tectonics that first revealed the dynamic nature of planet Earth and triggered the geoscience revolution that has profoundly changed human understanding of Earth's evolution. Now, the Fourth Industrial Revolution is around the corner. What, exactly, are the frontier scientific issues in the geosciences? The nations with leading science and technology, such as the USA and Europe, have proposed their own developmental visions called "Future Earth" and "Earth Survival Plan," respectively. Besides, six global organizations including the European Geosciences Union (EGU) and the American Geophysical Union (AGU) have jointly issued a "Declaration on the Importance of Earth Science Expertise to Address Global Social Challenges." In addition, Chinese Earth scientists have presented that "The nature and evolution of habitability of the Earth" is the frontier scientific issue in the twenty-first century.

Venus, Earth, and Mars are all located in the "habitable zone" in the solar system, as defined by astronomers, yet only Earth among them has evolved into a suitable planet for life. From the perspective of human habitability, and for sustaining life on Earth in general, the planet we live on requires the following characteristics: (1) clean air, and water, along with sufficient nutrients, (2) healthy and comfortable living space, (3) self-repairing capability to mitigate the consequences of natural disasters within and beyond the Earth system, (4) the ability to adjust to climatic, ecological, and environmental changes, and (5) abundant resources, particularly renewable resources, to sustain life. To cope with the challenges presented by both natural and human-impacted changes, it is necessary to make full use of emerging technologies, big data, and comprehensive observational data to improve human beings' insight and ability to assess the evolution of the habitability of Earth. This constantly evolving body of knowledge is changing current research paradigms and supporting transformative technologies that facilitate interdisciplinary integration which leads to new approaches for research regarding the habitability of the Earth.

Over the past decade, in support of scientists, the National Natural Science Foundation of China has funded a series of major research projects, such as "Destruction of the North China Craton," and the studies of Earth's evolution. Based on relevant cutting-edge research projects, this book integrates the knowledge and intelligence of Chinese geoscientists and discusses future trends of our environment along with strategies to overcome upcoming challenges. To do this, we first introduce the fundamental properties, state, and evolutionary laws of the universe and the Earth. Then, we discuss the scientific, technological, and practical issues that we face as we seek to protect the "Habitability of the Earth," offering a recommendation to expand the fields available for interdisciplinary studies. The book is divided into five sections. The first section provides an "overview" of the viewpoint that "mastering the developmental laws of the Earth acting on mankind, guiding human survival activities, and eventually realizing the sustainable development, is a crucial scientific proposition for future Earth science research." This broad statement is assessed from the aspects of scientific research on deep Earth, the deep sea, deep space, and terrestrial systems, and policies for research. The second section entitled "Scientific Vision" explores the scientific connotations of the "Habitability of the Earth" for expanding scientific horizons. The third section, "Scientific Breakthrough," describes the fundamental theoretical issues in research on the "Habitability of the Earth" from the perspective of national needs. The fourth section, "Technological Support," describes the transformative technologies in support of research on the "Habitability of the Earth." The fifth section deciphers ideas and makes suggestions for how to carry out the research mentioned above, including platform setup and data sharing, interdisciplinary and collaborative research, international cooperation and exchanges, etc.

This *Development strategy of Earth Science from 2021 to 2030*" is drafted on the basis of two years of group research on perspectives for the development of the Earth sciences in the next decade. The objective of the work is for facilitating sustainable development of Earth science in China. This report is a debut for the Chinese Earth science community, and we welcome criticism and feedback from the entire geoscience community and society at large. The research was funded solely by the Division of Earth Sciences of the National Natural Science Foundation of China, while the contents of the book only represent the point of view of the authors. Here, I would like to express my gratitude to the individuals who have participated and contributed to this development strategy.

2021   Rixiang Zhu
                                                                            Institute of Geology and Geophysics
                                                                                      Chinese Academy of Sciences
                                                                    Academician of the Chinese Academy of Sciences
                                                                                                      Beijing, China

The original version of the book was revised: The institutional author name has been updated. The correction to this book is available at https://doi.org/10.1007/978-981-19-2783-6_6

Introduction

This book explores the theme of the "Habitability of the Earth," highlighting deep Earth, deep sea, and deep space, as well as the surface systems which have close contact to human life. The content of this book is based on the latest research by Chinese geoscientists working at the frontiers of related fields in order to form some preliminary ideas on the scientific and technological research supporting "Habitability of the Earth" and to lead future studies in this theme. This book is the first "white paper" conducted by Chinese Earth scientists. It is recommended to geoscience researchers and students for reference.

Contents

1 Overview 1
   1.1   Deep Processes Control Earth's Habitability 2
   1.2   Understanding the Main Factors of Earth Habitability
        by Ocean Exploration 3
   1.3   Interaction Between Earth's Endogenic and Exogenic
        Processes from a Deep Space Perspective 5
   1.4   Earth System Science and Earth Habitability 6
   1.5   "Ecosystems" of Innovation 7
   Reference 9

2 Scientific Perspectives: Challenges for Human Cognition 11
   2.1   Early Earth 11
   2.2   Effect of Deep Dynamic Processes on Earth's Habitability 13
        2.2.1   Materials Circulation and Deep Earth Structure 13
        2.2.2   New Physical Chemistry of the Deep Lower Mantle
              and Deep Earth Engine 14
   2.3   Influence of Important Events on Earth Habitability 16
        2.3.1   Deep Earth Processes and Constant Temperature
              Mechanisms of Earth's Climate 16
        2.3.2   Why Does Plate Tectonics Occur Only on Earth
              and How Does It Affect the Habitable Environment? 17
        2.3.3   Influence of Significant Geological Events
              on the Environment and the Evolution of Life 18
   2.4   Earth's Atmosphere and Climate Change 19
        2.4.1   Oceanic Heat Absorption and Carbon Sequestration,
              and Their Role in Climate Change 19
        2.4.2   Climate Change in Extremely Hot Geological Periods 20
   2.5   Interactions Between the Oceans and the Earth's Interior 21
        2.5.1   Lithosphere Structure and Composition of Oceanic
              Plates 21
        2.5.2   Driving Forces of Oceanic Plate Movements 22

2.5.3   Deep-Water Circulation and Sea Level Fluctuation   23
2.6   Evolution of the Oceans and the Origins of Life   23
2.6.1   Extreme Oceanic Life Processes and Origins of Life   23
2.6.2   Evolution of Marine Life and Its Adaptability
to the Environment   25
2.6.3   Role of Marine Life in Earth Evolution   25
2.7   Multi-sphere and Multi-scale Interaction Processes
and Physical Mechanisms of the Sun and Earth
in a Deep-Space Environment   27
2.7.1   Sun and Earth Interaction   27
2.7.2   Geological Activity and the Formation Mechanisms
of Icy Celestial Bodies   27
2.7.3   Study and Evaluation of Small Extraterrestrial Celestial
Bodies, and Prediction and Prevention of Earth Impacts ...   28
2.8   Detection of Habitable Extrasolar Planets   28
2.9   The Human-Earth System and Sustainable Development   29
2.10   Global Environmental Change and the Evolution of Biology
and Human Culture   31
2.11   Sphere Interactions and Earth System Processes   32
References   34

3   **Basic Scientific Issues Relating to Earth Habitability**   37
3.1   Important Issues in Resources and Energy Security   37
3.1.1   Oil and Gas Resources   37
3.1.2   Mineral Resources   39
3.1.3   Surface Water Resources   40
3.2   Deep-Sea Resources, Energy Potential, and Marine Security   41
3.2.1   Deep-Sea Energy   41
3.2.2   Deep-Sea Mineral Resources   41
3.2.3   Deep-Sea Biological Resources   42
3.2.4   Marine Security   43
3.3   Deep-Space Resources and Deep-Space Economy   44
3.3.1   Exploration, Development, and Utilization of Space
Resources   44
3.3.2   Utilization and Transmission Technology
for Deep-Space Solar Energy   44
3.3.3   Multi-scenario Economic Development in Space   45
3.3.4   Theories and Technologies for Human Intervention
in the Trajectories of Dangerous Small Celestial Bodies ...   45
3.4   Mechanisms, Prediction, and Prevention of Natural Disasters   46
3.4.1   Earthquake Prevention and Mitigation of the Effects
of Strong Continental Earthquakes   46
3.4.2   Observation, Mechanism and Possible Control
of Artificially Induced Earthquakes   47
3.4.3   Marine Hazard Forecast and Prediction   48

3.5   Ecological Safety 48
      3.5.1   Ecosystem Structures and Processes 48
      3.5.2   Soil Health 49
      3.5.3   Environmental Pollution Control 50
3.6   The Carbon Cycle and Carbon Neutrality 50
      3.6.1   Cross-Sphere and Multi-scale Processes
              and Mechanisms of the Carbon Cycle and Their
              Relationships with the Climate System 52
      3.6.2   Budgets, Reservoir Capacity, Uncertainty,
              and Evolution Trends of Carbon in Terrestrial, Oceanic,
              and Land-Sea Coupled Systems 52
      3.6.3   Impact of Carbon Neutrality on the Coupling System
              Between Carbon Cycling and Climate 53
      3.6.4   Scientific and Technological Basis for Negative
              Emissions Technology 53
References 54

4   Scientific and Technological Support: Fundamental Theoretical
    Issues with Revolutionary Technologies 57
4.1   Geophysical Exploration Technologies for Deep Earth 57
      4.1.1   Theories and Technologies for Seismic Surveys 57
      4.1.2   Quantum Sensing and Deep Geophysics 58
4.2   Deep-Earth Geochemical Tracker and High-Precision Dating
      Techniques 59
      4.2.1   Geochemical Technologies for Tracing Early-Stage
              Earth Evolution and Core-Mantle Differentiation 60
      4.2.2   Index System of Earth Habitability Elements 60
      4.2.3   High-Precision Dating Technique 61
      4.2.4   HPHT Experiments and Computing Simulation
              Techniques 62
4.3   Deep-Sea Observation and Survey 63
      4.3.1   Detection Technologies for Ocean Laser Profile 63
      4.3.2   Detection Technologies for Marine Neutrino 63
      4.3.3   Deep-Sea and Transoceanic Communication
              Technologies 63
      4.3.4   Underwater Observation and Survey 64
      4.3.5   Sea Floor In-Situ Surveys 64
4.4   Deep-Sea Mobility and Residence 65
4.5   Exoplanets Exploration 66
      4.5.1   Exploration of Atmospheres and Life on Exoplanets 68
      4.5.2   Climate Environment on Oceanic Planets 68
      4.5.3   Ocean Currents and Heat Transportation
              on Lava Planets 69

4.6   Infrastructure Framework and Theories of Positioning,
       Navigation, and Timing (PNT) Services 69
       4.6.1   Collaborative PNT Systems in High, Medium, and Low
               Orbits 71
       4.6.2   Pulsar Space–Time Datum 72
       4.6.3   Marine PNT System 72
       4.6.4   Miniaturized and Chip-Size PNT 74
       4.6.5   Theory and Technology of Quantum Satellite
               Positioning 74
       4.6.6   Datum Measurement Technology for Optical Clocks
               and Elevations 75
       4.6.7   Technology and Fundamental Theory of Resilient PNT 77
4.7   Big Data and Artificial Intelligence 78
       4.7.1   Data Integration, Data Assimilation, and Knowledge
               Sharing 78
       4.7.2   Geoscientific Knowledge Graphs and Knowledge
               Engines 79
       4.7.3   Deep Machine Learning and Complex Artificial
               Intelligence 81
       4.7.4   Data-Driven Geoscientific Research Paradigms
               Transformation 82
References 84

5 Realization: Intersectionality, Integration, Collaboration,
   and Cooperation 87
   5.1   Platform Construction and Data Sharing 87
   5.2   Interdisciplinary and Collaborative Research 88
   5.3   International Collaboration and Exchanges 88

Correction to: Past, Present and Future of a Habitable Earth C1

Postscript 91

Chapter 1
Overview

With basic needs for survival being satisfied, prosperous humans began to explore the mysteries of the Universe and the Earth where life was originated and evolved ever since. The celestial mysteries of sunrise and sunset, the movements of planets and their satellites, and discoveries and adventures of travelers and explorers motivated the curiosity and research about the Earth and its place in the universe. However, the progress of scientific research and development in today's world is determined no longer just by curiosity and wonder, but mainly by the imperative of living activities and technological advances. From the Industrial Revolution in the eighteenth century to the present digital age, geoscience has progressed from academic research on understanding basic physical laws, chemical changes and evolutionary processes, to multi-disciplinary pure and applied research in geology, geophysics, geochemistry, atmosphere and ocean, and comparative planetology, etc., driven by the need to acquire resources and to avert natural disasters. In the 1960s, geoscientists studied continental drift, volcanic activity, and sea-floor spreading from a global perspective, and established a new view of the Earth—plate tectonics theory. Also in the 1960s, mankind successfully broke away from the gravity of Earth and set foot on the Moon. The research on global change that began in the 1980s extended the study of the human living environment to span the entire Earth system, combining the study of Earth's interior with research on surface processes and terrestrial space, and from tracing the cosmic explosion to human intelligence which eventually formed the new view of point of "Earth system science".

Today, Earth science research has experienced revolutions in computer and information technology and is equipped by new perceptions from technology and digital processing technology to be capable of understanding of all aspects of the entire Earth and even the universe system from observations from an unprecedentedly powerful observation network on land, at sea, and in the air and space. At the same time, human exploitation and utilization of Earth's resources and other impacts on the Earth system are increasing exponentially, posing enormous questions for the future. Is mankind heading towards the self-destruction of resource exhaustion and

© Science Press 2022, corrected publication 2022
The Research Group on Development Strategy of Earth Science in China,
Past, Present and Future of a Habitable Earth, SpringerBriefs in Earth System Sciences,
https://doi.org/10.1007/978-981-19-2783-6_1

the deterioration of the living environment, or mastering the development law of the Earth's role in human beings, guiding mankind's survival activities, and achieving sustainable development? The latter is an important scientific proposition for Earth science research in the future.

With the theme of "Earth habitability", this book highlights the deep Earth, deep sea, and deep space, as well as the surface systems closely related to the human living environment. We have studied the latest research, drawn on the great achievements of Chinese geoscientists working at the frontiers of related fields, based on expanding our scientific horizons, to form some forward-looking ideas and proposals for future research on the scientific and technological support, and the practical measures, required to ensure "Earth habitability".

1.1 Deep Processes Control Earth's Habitability

Through billions of years of evolution, the Earth has gradually evolved from a relatively uniform and comparatively hot planet into a vibrant and habitable world with a robust sphere-layered structure. However, Earth's habitability is not an innate quality of the planet. The Big Bang, which eventually led to star formation, solar system formation, and the birth of Earth, resulted in the formation of a magma ocean covering the entire Earth. Meanwhile, the strong reducing environment created a dense, toxic atmosphere predominantly composed of carbon dioxide, methane, hydrogen, ammonia, and water vapor. How did the Earth then evolve into the planet we know today, with a mild, habitable environment that has supported and promoted the emergence of life? How has the Earth developed to possess strong self-regulation and repair capacity to maintain a relatively stable livable environment? How does it produce the resources and energy required for human survival and sustainable development? These questions represent key issues of modern Earth system science and the answers can be found through understanding interactions between Earth's spheres—atmosphere, biosphere, lithosphere, hydrosphere, etc.—and their effects on resources and environment. Studying the developmental process, key control factors and regulation mechanisms of Earth's habitability are fundamental for predicting the future of Earth, and is a vital basis for discovering more resources and energy as well as maintaining the survival of human beings and sustainable social development.

Earth's habitability, in particular its deep controlling mechanisms, remains to be explored. Current understanding of the evolution of the habitable Earth can be divided into two models: the "quasi steady-state model" and the "catastrophe model". The latter model encompasses well-known major geological phenomena, such as the Great Oxidation Event, the 'Snowball Earth', Oceanic Anoxic Events, and Biological Extinctions, all of which significantly shaped the Earth system. Of particular note is the occurrence of the Snowball Earth before the Cambrian "explosion" of animal life, and correlations between oceanic anoxia/rapid increases in oxygen in deep Earth and rising oxygen in the atmosphere, and between assembly/dispersal of supercontinents and Earth's habitability.

What is the nature of these correlations? What causes putative biological extinctions and the emergence of new life forms? What triggers disastrous environmental and climatic change? What is the origin of oxygen which became considerably enriched during the Great Oxidation Event? What was responsible for extensive melting of the deep mantle and the formation of flood basalts and large igneous provinces? What disturbed normal mantle convection and triggered the assembly and dispersal of supercontinents? Why are there two dynamic processes in Earth's interior: frequent reversals of geomagnetic polarity, and the long-term superchrons? Understanding the intrinsic linkage between these seemingly independent major events in the various spheres may provide the key to understanding the operational mechanisms of the entire Earth system. Previous research has generally focused on individual events. Although some advances have been achieved, most hypotheses tend to attribute the causation of events to external factors. For example, a mass extinction event has been attributed to an asteroid impact, subsequent sea-level fluctuations, etc., but the effect of the Earth's interior dynamic processes has largely been underestimated. Consequently, hypotheses and theories are often contradictory. The atmosphere, biosphere, hydrosphere, and lithosphere account for only a small proportion of the Earth's total volume, whereas the vast interior of the Earth is largely unknown. Carbon, hydrogen, oxygen, and sulfur are essential for life. However, their reservoirs on Earth's surface only account for less than 1% of those of Earth. The vast majority of the Earth's carbon is held in the planet's deep interior (Plank and Manning 2019). Knowledge of deep carbon cycling will permit a better understanding of the current debate on carbon emissions in the international community, which is largely focused on surface and atmospheric carbon cycles, and the evolution of Earth's climate systems. It is therefore of paramount importance to investigate how deep Earth dynamic processes modulate the operation of the Earth system and regulate Earth's habitability. A universal vision is also required if ones wish to better understand the Earth's past, present, and future.

1.2 Understanding the Main Factors of Earth Habitability by Ocean Exploration

The deep ocean is an important regulator of the Earth's climate, controlling the cycles of carbon, hydrogen, oxygen, sulfur, and other materials of the whole Earth system. An in-depth understanding of the ocean and related seafloor dynamic processes, and their energy and material cycles, will assist in revealing the operations allowing Earth habitability and in balancing future developmental needs and carbon emissions. The ocean is also the largest ecosystem on Earth. Understanding the processes and laws of this "blue" living system, and rational development and protection of blue biological resources are significant strategic needs to support the sustainable development of human society. As the "blood" of the Earth, the oceans are vital links connecting all spheres of the Earth system. The oceans are vast and deep, and remain largely mysterious, with less than 5% explored by mankind so far. China is a latecomer to marine science, so it is imperative that in the future we should undertake major scientific projects such as the "Transparent Ocean", the "Deep-sea Base Stations", and the "Ocean Drilling Project (ODP)" focused on the exploration concept of "deep-sea entry, deep-sea exploration, and deep-sea development". This will encourage and assist scientists of all nations to develop new theories and technologies in the fields of marine and polar research.

The seabed is a lightless, but vibrant world. In the past 50 years, one of the most important scientific achievements in deep-sea research is the revelation that life does not necessarily depend on photosynthesis, which may open an entirely new window into the origins of life on Earth. Significant discoveries from seafloor hydrothermal vents and fluids, particularly low-temperature alkaline hydrothermal methane, cold springs, and deep-sea dark-life systems, have been providing clues about the transformation of life from inorganic to organic matter. Geoscientists, chemists, biologists, and others are working together to illuminate the genesis of methane and various protein substances in the eruptive processes of submarine volcanoes, helping unravel the mystery of the origin and suitability of life on Earth.

With continuous population growth, the human race generates ever more critical needs for energy, mineral resources, water resources, and even spatial resources. As we relentlessly exploit Earth's resources to satisfy these needs, the marine ecological environment and Earth's climate system, which are inextricably linked to human comfort and survival, are facing new levels of threat and uncertainty. To address this, we should analyze the influence of fluid–solid interactions in the Earth system. Seawater affects the physical properties and chemical compositions of the ocean plates through seabed alteration, sedimentary processes, hydrothermal activity, etc. Plate subduction draws surface fluids down to the deep Earth, impacting magmatism, exchanges of material between the spheres, and even the entire evolution of the Earth. The partial melting of the mantle induced by water entering the Earth's interior may have contributed to the high oxygen fugacity in the asthenosphere and lower mantle. These rigorous scientific studies have been instrumental in identifying and understanding the forces driving plate movements and the genesis of Earth's Great Oxidation Event.

For the future, the mechanisms of activity in deep molten melts/fluids, such as the water and carbon cycles driven by plate subduction, should be at the heart of efforts to understand the wider factors affecting Earth habitability. Multi-sphere coupled research, with the oceans as the core, will be essential for understanding and managing the hydrosphere, and will be intricately connected to the sustainable development of human society.

1.3 Interaction Between Earth's Endogenic and Exogenic Processes from a Deep Space Perspective

The Earth is the only planet in the solar system that is known to support life. Earth habitability correlates not only with the Earth system itself, but also with the activity of the Sun, the rest of the solar system, and even more distant celestial bodies. The 'habitable zone' of the solar system, defined by astronomers according to distance from the Sun, contains the three Earth-like—or terrestrial—planets, Venus, Earth and Mars. The terrestrial planets were formed 4.6 billion years ago, coalescing from rings of debris around the sun during the early evolution of the solar system. Nevertheless, about 4–3.5 billion years ago, the evolutionary paths of Venus, Earth and Mars began to diverge considerably. Today, Venus has ninety-two times the atmospheric pressure of the Earth, an atmosphere composed almost entirely of carbon dioxide (96.5%), and surface temperatures that reach 460 °C, which is typical of a runaway greenhouse effect. Mars has an atmospheric density less than 1% of the Earth's atmosphere, and the atmosphere is 95% carbon dioxide. The surface temperature is as low as –63 °C, which represents a runaway icehouse effect. Of the three, only Earth has evolved into a habitable planet. Thus, the formation of a habitable planetary environment is related to other internal factors in addition to the planet's distance from its star, with the endogenic dynamic processes of the planet apparently being the dominant factors. Mars lost its magnetic field about 3.5 billion years ago, and only residual crustal magnetic fields remain. The absence of global magnetic field protection may be responsible for the loss of almost all of Mars' original atmosphere and liquid water. The mass of Venus is similar to that of the Earth, but evidence of frequent, intense volcanic activity indicates primarily vertical tectonic movement, with nothing much resembling Earth's unique tectonic plates drifting horizontally. In contrast, complex vertical and horizontal movements occur simultaneously in the Earth's endogenic dynamics, with differential rotational motion creating a 'dynamo' effect that generates the planet's strong global dipolar magnetic field, which is an essential element in Earth habitability. What caused these striking evolutionary differences in the evolutions of Mars, Venus, and Earth? Comparative planetology should address issues such as the spatial and temporal evolution of the inner and outer spheres of the terrestrial planets, the roles of water and volatile substances, tectonic movements, and the role of magnetic fields. Until now, we have focused principally on Mars, various icy satellites of the outer gas planets, and even exoplanets in our quest for extraterrestrial life. This search addresses the question of whether humanity is alone in the universe, and encompasses our desire, and perhaps need, to expand the future living space for human beings beyond our single planet.

From the perspective of deep space, Earth habitability should be investigated from five perspectives: the 'solar-terrestrial panorama', planetary genealogy, exoplanet research, Earth defense research, and space–time benchmarking. The solar-terrestrial panorama focuses on the impact of solar activity on Earth habitability, and plane-tary genealogy examines the formation and evolution of Earth habitability by using comparative planetology to compare Earth with the other planets and bodies in the solar system. Exoplanet research seeks to understand the past, present, and future of the Earth through the detection and study of planets orbiting around other stars. Earth defense research considers the potential threats posed to human society by deep space events, such as impacts from small celestial bodies and also the monumental forces generated by extreme celestial bodies, exploring methods and countermeasures for disaster prevention and alleviation. Space–time benchmarking constructs the theories, technologies and systems for positioning, navigation, and timing (PNT) essential for deep space activity. Taking the Earth as the core, we will address major frontier scientific issues such as the Earth habitability's space environment, the origin and evolution of the other planets and their satellites in the solar system, and the habitability of exoplanets. We aim to reveal the entire story of the emergence and development of Earth habitability and contribute to expanding the space and resources available for human survival and social development. In doing so, a better understanding of the solar system and the universe will emerge, which will greatly assist us in moving towards a more far-ranging habitable universe and, perhaps, discovering a new home for mankind.

1.4 Earth System Science and Earth Habitability

The interactions between human activities and the natural processes in the Earth system is leading to a series of major challenges with regards to the human living environment, such as water scarcity, land degradation, biodiversity loss, and more severe environmental pollution. Resolving these issues necessities to understand the Earth system processes well beyond the societal timescale.

For instance, carbon emitted into the atmosphere by humans has interacted with carbon reservoirs in the ocean, soil, and ecosystem, with quite distinct life cycles. Tracking the global carbon cycle needs to consider the cross-spheric interactions of much wider spatial–temporal scales in integrating both anthropogenic and natural processes. The lack of a detailed understanding on the Earth's past, including the role of the Earth's interior and the interactions between the different spheres, would seriously handicap the accurate prediction of the future environment and its impact on the human society.

From this point of view, one of the major challenges for the future Earth System Science is to break the timescale barriers for comprehensively understanding the cross-scale interactions between the natural and human-induced processes, ideally from the human society timescale to the geological timescales.

Environmental change and biological evolution, as well as the evolution of civilization, combine to alter the Earth's biosphere. Each of these factors has their own internal laws, interactions, and coordinated development, collectively constituting a complete history of Earth's biology. It is well-known that the evolution of life is influenced by the Earth system dynamics. On the other hand, the evolution of life is also shaping the Earth habitability. The co-evolution of the Earth's environment and biology is therefore of great significance for understanding the Earth habitability in the future.

Natural and human-induced disasters and environmental pollution are increasingly challenging the sustainability of the habitable Earth. Natural disasters span a large spatiotemporal scale of formation and evolution, representing a typical giant, non-linear, complex system that can cause abrupt change in the whole or regional Earth system. In contrast, human-induced disasters are relatively more predictable in spatial and temporal range. The environmental pollution is now considered to be a great threat to the habitable Earth. In recent years, epidemiological surveys have shown that environmental factors now exceeding genetic factors have become the most dangerous factors affecting human health. For example, the global pandemic of COVID-19 is currently urging geoscientists to pay more attention to the Geo-Health (Earth and life health), as existing knowledge systems and technical capabilities are inadequate to guarantee the harmonious development of mankind and a habitable Earth.

1.5 "Ecosystems" of Innovation

In order for breakthroughs to occur in both science and technology on the world's frontiers, Earth science research needs to focus more on basic and fundamental research, in particular on those fundamental theories which can be then transformed into new technologies. To achieve these goals, it is essential to optimize policies that facilitate the establishment of shared platforms, interdisciplinary research teams, collaborative research projects, and promote outreach and nourish the next generation of talent and new-levels of international cooperation and exchange.

In retrospect, the history of modern science and technological development, beginning with the first industrial revolution which originated in Great Britain in the 1750s, was largely based on scientific theories such as classical mechanics, and marked by the development of the steam engine. This was a huge leap from reliance on wind, water, and animal power. Industries such as textiles, coal, metallurgy, and machinery manufacturing blossomed, and capitalist productivity was fundamentally improved, encapsulated in the brilliance of Victorian Britain—the world's first pre-eminent industrial power. In the 1860s, the second industrial revolution emerged rapidly and simultaneously in Europe, the United States, and Japan. Science played an ever more prominent role in supporting the development of productivity. A series of major scientific and technological innovations brought enormous changes to society, and mankind entered the electrical age. The third industrial revolution occurred primarily in the United States after the Second World War. It was marked by nuclear energy, aerospace technology, and information technology. It gave birth to new fields such as nuclear power, computers, aerospace, and a seemingly continuous stream of new materials and processes such as plastics and automation. Mankind entered the information age. Science and technology became ever more closely interconnected, accelerating the transformation of science and technology into direct productivity and boosting the United States to become a world superpower. The fourth industrial revolution, which began gradually in the twenty-first century, is chiefly based on cyber-physical systems, with the Internet and the Internet of Things, artificial intelligence, and big data as the core, and with the development of graphene, genetic engineering, virtual reality, quantum information technology, controllable nuclear fusion, clean energy, and biotechnology as feature breakthroughs. However, it can be argued that the current apparent technological revolution is simply the aftermath of the third industrial revolution. In fact, there have been almost no truly groundbreaking scientific discoveries over the last 20 years, and there have been no major breakthroughs in basic theories of the kind represented by relativity and quantum mechanics in the early twentieth century. The world's scientific and social practices in the first 20 years of the twenty-first century have only served to prove that basic research is the source of all scientific and technological innovation and a strategic cornerstone of building a scientific and technological world superpower. Although China has made unprecedented progress in basic research, our original contribution to the knowledge base of modern science and technology is still small, and consequently we lack leading core technologies. If China intends to make significant accomplishments in the fourth industrial revolution, we must enhance our efforts in basic scientific research and promote technological breakthroughs in our own right. Basic research is characterized by high investment and slow returns, and is difficult to evaluate quantitatively. Nevertheless, it determines a country's ability to achieve original innovation in science and technology. The implementation of the Morrill Act in 1862 greatly strengthened basic research in the United States, and contributed materially to the eventual rise of the United States as a world power in the early twentieth century. Our scientists should bolster their self-confidence and dare to face the most challenging cutting-edge scientific questions. They should be encouraged to put forward original theories, work hard on achieving groundbreaking technological innovation, and strive to achieve leapfrog development in crucial scientific and technological fields. Ultimately, we must make every possible effort to keep up with, and ultimately lead, the development of the fourth industrial revolution.

Science is a branch of culture. Scientific innovation requires a sympathetic cultural atmosphere, which is a social and public creation. Just like public transportation, the cultural atmosphere also needs the guidance of government policy. In the past few years, China has fostered a cultural atmosphere that measures scientific achievements largely according to the numbers and citation rates of academic papers published by individuals and institutions, forcing the most innovative young people to spend valuable time simply tracking and emulating the progress of international scientific research. Unfortunately, most current research only provides data support for the established theories of western scientists, and suggests a lack of critical thought. As a result, they lack original innovation and cannot meet the needs of the nation. Habitually following research trends in developed countries has made strategic planning in China simply an extension of international science strategies. As a result, our scientific research is not well integrated with the general trend of national development, and is out of step with the long-term goals and social needs of the Chinese people. For example, Chinese scientists are keen to engage in the study of nanomaterials rather than steel materials, because it is much easier to find publishers for nanomaterials research papers. This lack of basic steel research has led to the ludicrous situation that China is simultaneously the world's largest steel exporter and the world's largest importer of steel products. In the field of Earth science, we should strive to contribute wisdom to major scientific issues relating to the deep Earth, deep sea, deep space, and the Earth system to ensure sustainable development in crucial areas such as resources, energy, and ecological civilization. We need to have a deep understanding of the past, present, and future of a habitable Earth.

Tolerance of failure is another vital element of a healthy culture of innovation. At present, the science and technology management system and the mass media in China only welcomes successes in scientific and technological innovation. In fact, the eagerness for 'quick wins' and instant benefits is the greatest obstacle to true scientific and technological innovation. It often, if not usually, takes decades for an original, innovative theory to produce something of social and economic value. Therefore, to build an innovative nation, it is not enough to encourage scientists to be dedicated to scientific research. The government should also create a genuinely tolerant cultural atmosphere for scientific and technological innovation. As early as the sixteenth century, King Frederick II of Denmark gave the astronomer Tycho Brahe the entire island of Hven to build an observatory, and provided him with abundant funding for scientific research and living expenses throughout his life. There, Brahe carried out groundbreaking and historically pivotal astronomical work. Fortunately, Frederick II never asked or expected Brahe to be economically productive for his research. We must reflect seriously on how researchers can achieve truly original innovation in the current commercially driven international environment for scientific research.

Reference

Plank T, Manning CE (2019) Subducting carbon. Nature 574(7778):343–352

Chapter 2
Scientific Perspectives: Challenges for Human Cognition

2.1 Early Earth

One of key issues in the study of Earth's habitability concerns the origin and preservation of volatiles, which are related either to Earth accretion processes associated with some special giant impacts or to solidification of magma ocean and vertical tectonics, which were responsible for volatile transport and distribution. Obviously, all these processes are associated with the evolution of early Earth (Armstrong et al. 2019), understanding of which requires following investigations in the future: ① Earth accretion processes (material types and acquisition of volatile substances); ② the number and pattern of giant impacts on the Earth and their influence on the preservation of different types of volatiles; ③ the connections between giant impact events and the present core-mantle boundary structure, and how did giant impacts cause the preservation of some primodial components in the lower mantle; ④ how did the core-mantle differentiation form Earth's layered structure; ⑤ the scale of the latest magma ocean and its solidification, and the probability of the existence of a deep magma ocean; ⑥ precise dating of major geologic events; ⑦ primitive core constituents and the onset of the Earth magnetic field, etc.

One of significant difficulties in carrying out this research is the almost complete lack of geologic samples. Although ~4.4 Ga zircons and primordial materials as old as the Earth itself residing at the core-mantle boundary (brought to the surface by mantle plumes, Mundl-Petermeier et al. 2020) have recently been discovered, they are nevertheless indirect samples in the sense that they have been displaced from their genetic environment and therefore cannot provide a full range of information. Earth's original appearance and constitution can only be inferred from these indirect ancient samples using isotopic and elemental geochemical methods. Meanwhile, high-temperature and high-pressure experiments, Earth dynamics calculation, comparative planetology, and planetary accretion dynamics should be conducted to simulate early Earth evolution (Fig. 2.1). Future Earth scientists must integrate these study methods and develop new research paradigms.

© Science Press 2022, corrected publication 2022
The Research Group on Development Strategy of Earth Science in China,
Past, Present and Future of a Habitable Earth, SpringerBriefs in Earth System Sciences,
https://doi.org/10.1007/978-981-19-2783-6_2

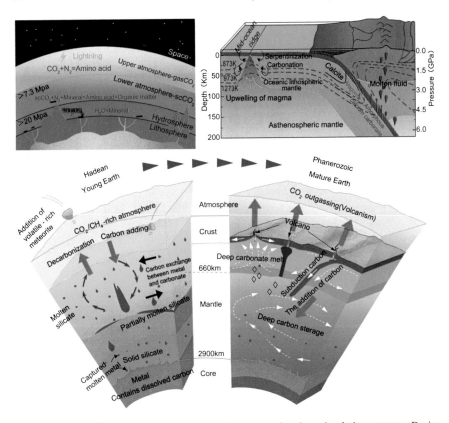

Fig. 2.1 Diagram illustrating showing deep-Earth water and carbon circulation patterns. During magma ocean

The Earth's magnetic field is one of major factors in providing the necessary conditions for the formation and development of Earth's habitability. Its formation is highly relevant to understandings of the Earth's formation and evolution, deep interior dynamics and space environment. Earth's magnetic field is generated by the 'dynamo' induced by motion in the liquid outer core. The 'dynamo' transforms some of the kinetic energy of fluid motion into electromagnetic energy by MHD (magnetohydrodynamics) and electromagnetic induction. Autogenesis and mainte-nance of Earth's magnetic field involve planetary rotation, the presence of conductive fluids, and sufficient heat energy (Olson et al. 2018). The existence of a solid inner core, around which molten conductive fluids rotate, is also a likely contributing factor in generating a strong magnetic field. The key issues for research on Earth's magnetic field include the mechanisms of formation and operation. Although impor-tant advances have been made in this field in recent years, the dynamic parameters required for the maintenance of the geomagnetic dynamo remain unclear.

During magma ocean crystallization in early Earth, volatiles (water and carbon dioxide) rose to the surface. At that time, there was over 100 atmospheric pressure of

CO_2, several hundreds of atmospheric pressure of water vapor, and a little nitrogen in the atmosphere. A supercritical CO_2 layer between the primitive ocean and the atmosphere might be the key to the origin of life (Zhang et al. 2020). When plate tectonics operate on Earth, plate subduction and volcanic activity are the main cycling pathways of material circulation between Earth's interior and exterior (Schmidt and Poli 2014; Dasgupta 2013; Hou et al. 2019; Zhang et al. 2020).

2.2 Effect of Deep Dynamic Processes on Earth's Habitability

2.2.1 Materials Circulation and Deep Earth Structure

Materials and energy exchanges between Earth's different spheres are essential to the 'dynamic Earth' and are also crucial processes in making our planet habitable and favorable for the birth of life. Spreading mid-ocean ridges, plate subduction and mantle plumes are the principal pathways for materials and energy exchanges between the solid Earth's different spheres. These apparently independent dynamic processes are indeed mutually and closely related.

Oceanic lithosphere forms at mid-ocean ridges through mantle partial melting, representing one of principal sources of mantle cooling. Thermal fluid-rock reaction at in these areas not only alters raw materials supplied to 'subduction factory', but also affects sea water composition and submarine life processes. Plate subduction zones are vital sites for materials and energy exchange, fluid–solid couplage and Earth's sphere interactions. Subduction acts an efficient circulation pathway in linking deep interior processes and surface processes, governing both the interior and exterior evolutionary processes of the globe. Dehydration of subducting slab and upward transfer of water trigger melting of the overlying mantle wedge and the formation of island arc or continental arc magmatic rocks, providing materials for the reworking of continental crust. In this sense, the subduction zone is considered as graveyard of subducting oceanic crust and birth site of new continental crust.

A great amount of slab materials is transported to the deep mantle by subduction, creating the heterogeneity of the mantle in terms of lithology, elements, isotopes, and volatiles (H_2O, CO_2, etc.). Mantle heterogeneity would in turn induce mantle convection and magmatism, exerting significant influence on the formation and evolution of the oceanic and continental lithosphere, enrichment of mineral resources in the shallow crust and changes to habitable environment on the surface. The detention of subducted plates at the core-mantle boundary may have resulted in the seismically detected Large low-shear velocity provinces (LLSVP) and ultra-low velocity zones (ULVZ) (McNamara 2019; White 2015).

The total quantity of all the volatiles transported in subduction zones exceeds considerably those of respective components on the surface. Elements with variable valency such as C, H, O, S, etc. are responsible for the variation of redox state of earth's interior and have an enormous influence on the property of the Earth's interior and geodynamic processes. Subducted slab and their derivatives of dehydration and melting are present at different depths of the mantle. Time-evolution of these components and their interactions with ambient mantle ultimately generated distinct types of mantle end-members (EM1, EM2, HIMU, FOZO, etc.).

In order to understand materials and energy exchange between the different Earth spheres and to elucidate their roles in the formation and evolution of habitable Earth, investigations from the Earth system viewpoints need to be carried out on the structure and origin of Earth's different spheres, compositions and origin of diverse mantle end members, and the dynamic processes governing materials exchange and energy circulation in subduction zones. It is also important to quantitatively estimate transfer fluxes of key elements during major geological events. Finally, these new acquired data will be integrated to evaluate the effects of deep-Earth processes on changes in Earth's habitability.

2.2.2 New Physical Chemistry of the Deep Lower Mantle and Deep Earth Engine

From the deep crust to the middle of the lower mantle (about 1800 km), minerals and rocks go through several facies changes, but the physical and chemical principles defined under normal temperature and pressure do not vary. However, in the lower part of the lower mantle (below 1800 km), familiar physical and chemical rules appear to change, and the properties and chemical behavior of the elements alter beyond recognition. For example, the properties of iron become similar to those of magnesium and the properties of hydrogen to those of lithium. When both ends of a single system exist under fundamentally different physical and chemical conditions, potentially extreme variations can occur, which may be evidence of an as yet unknown dynamic engine for Earth's evolution (Mao et al. 2017). For instance, the current mainstream view is that the lower mantle (660 ~ 2900 km and 24 ~ 135 GPa) consists of bridgmanite and ferropericlase. Although it accounts for 55% of Earth's total volume, it is thought to be in simple composition. Whereas, breakthroughs in X-ray spectroscopy and synchrotronic radiation analysis have revealed Fe-paired magnetic rotation of mantle minerals and the decomposition of bridgmanite under super-high pressure, as well as iron peroxide synthesis which releases hydrogen and remains vast amounts of oxygen when hematite comes into contact with water (Hu et al. 2017, 2020). This suggests that iron peroxide is a considerable component of the core-mantle boundary (Liu et al. 2017). These new understandings demonstrate that HPHT conditions in the deep lower mantle can have a profound effect on mineral density, wave velocity, melting temperatures, conductivity, magnetic properties, rheological intensity, and other physical properties, as well as on the controlling dynamics of the mantle, the heterogeneity of lateral structures and components, and the evolution of deep Earth. These new clues and discoveries appear to confirm that the super-deep

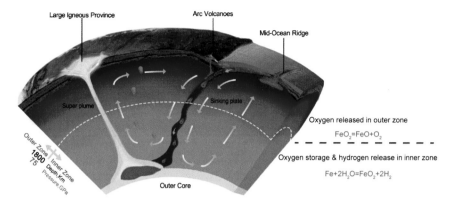

Fig. 2.2 Entirely different physical and chemical properties of the super-deep (>1800 km) and shallow mantle (amended by Mao and Mao 2020). The differences may relate to Earth's interior working mechanisms, and might also be the causes of important geologic events on the surface

mantle (>1800 km) and the shallow mantle are profoundly different (Fig. 2.2), which foreshadows another revolution in HPHT chemistry. Study of these questions will guide us in the construction of new theories and concepts.

Earth's evolutionary history began around 4.6 billion years ago (Ga). At 2.4–2.0 Ga a major oxidization event occurred and the diversity of life on Earth instantly began to increase exponentially. At about 550 Ma, the oxygen in the atmosphere abruptly increased again, coinciding with the advent of vertebrate lifeforms. This much is known, but what remains mysterious is the question of where the oxygen came from that drove these oxidization events. The conventional view has been that CO_2 on the surface was transformed into oxygen by cyanobacterial photosynthesis. This hypothesis matches the facts, as far as they are known, but there is no convincing geological evidence to support the hypothesis. More recent HPHT calculations and experiments have confirmed that, when subduction slabs subside at the core-mantle boundary, the water drawn down by subduction contacts with iron. Following a series of intermediary reactions that produce ferric oxide and FeH_2, the final product of this process is iron peroxide. A single subsiding tectonic plate could transport hundreds of millions of tons of water to the core-mantle boundary every year. Over billions of years, oxygen-rich blocks several thousands of meters thick probably accumulated at the bottom of the lower mantle (Mao and Mao 2020). Destabilization of these oxygen-rich layers could have released the oxygen into the atmosphere and generated the great oxidization event. The formation principle of oxygen-rich layers at the core-mantle boundary is that oxygen remains trapped there following release of hydrogen from the water drawn down by subduction, which suggests that the Earth's interior is a vast hydrogen-producing machine. It is already known and understood that hydrogen molecules tend to release across crystal lattices, but the precise patterns and processes of hydrogen volatilization need to be more closely studied in the future. First, the possibility should be explored of hydrogen combining with carbon to form abiogenetic hydrocarbons, or performing as a catalyst in the genesis of organic matter in the shallow crust. Second, the potential trapping of naturally occurring hydrogen in clay mineral layers should be considered. Third, the possibility of hydrogen recombining with oxygen to form water during geological uplifting should be examined. Fourth, we should determine how hydrogen could control the migration of volatile substances from the interior of the Earth in combination with nitrogen, sulfur, phosphorus, and halogens. These avenues of study will open significant new perspectives for the geosciences.

At present, our knowledge of the super-deep Earth is based mainly on high-temperature and high pressure (HPHT) measurements and experiments in geophysics and geochemistry. Previous studies have focused on synthesis of individual minerals under high pressure to measure their basic physical properties and thereby interpret deep seismic structures. Future research will focus on developing in-situ HPHT testing technologies, measuring deep rock mineral properties, exploring physical and chemical behavior, and controlling geophysical and geochemical phenomena in high pressure environments. This will cast light on the critical role of the super-deep regions in Earth's evolution from core to surface and from past to future, and will promote the re-evaluation of fundamental theories of a 4D Earth system.

2.3 Influence of Important Events on Earth Habitability

2.3.1 Deep Earth Processes and Constant Temperature Mechanisms of Earth's Climate

Since 4Ga, surface temperatures have been maintained in the narrow range that supports the existence of liquid water—an essential factor in the continuous evolution of life. By comparison, climate-controlling factors (for example, solar radiation, land-sea distribution, and CO_2 gas emissions from the mantle) have varied immensely. Deep tectonic movements have also developed from the vertical motion of the early Earth into a complex system of coexisting and simultaneous vertical and horizontal movements. The mechanism that has maintained stable surface temperatures over this immense period of time, while other environmental factors have fundamentally changed, is clearly one of the keys for deepening our understanding of the formation and development of Earth habitability.

The traditional view is that long-term surface temperature stability is related to a negative feedback mechanism formed by continental weathering, a process which has been called "continental weathering geological air conditioning". When the climate becomes warmer, the rate of atmospheric CO_2 adsorbed by silicate weathering increases, which leads to declining atmospheric CO_2 content, which in turn inhibits further temperature rises, and vice versa. Nevertheless, in the course of geological evolution, continental surfaces have been covered by sediments and granite that have undergone complete weathering cycles, with severe reduction in the efficiency of CO_2 adsorption by the weathered surface, weakening the capacity of the sediments for climate adjustment. The 'air conditioning' hypothesis is therefore not confirmed by the geological record. It is also difficult to explain why this mechanism should only have emerged on Earth but not on Venus or Mars, both of which are also within the Sun's habitable zone.

Unique deep Earth processes may be the key to understanding these questions. Perhaps large igneous provinces, rifting and expansion, and volcanoes in subduction zones provided a continuous supply of fresh basalt to the surface, which could have 'fueled' the process, since basalt weathering has extremely high CO_2 adsorption efficiency (Dessert et al. 2003), roughly one hundred times that of granite weathering. Exhausted volcanoes could also have continued to pump CO_2 from deep Earth to the surface, helping to keep the 'geological air conditioning' system working. Study of the synergetic evolution of deep Earth events (e.g., plate tectonics initiation, continental crust growth, supercontinent cycle) together with continental weathering, atmospheric CO_2 content, and climatic environment would be an effective way to verify the 'continental weathering geological air conditioning' theory. Research should concentrate on the reconstruction of deep Earth carbon cycle processes, mantle exhaustion history, and continental weathering history. Numerical simulation of deep crust-mantle circulation can be conducted by extracting chemical footprint information from weathered continental deposits and remnants of mantle exhaustion, which may offer significant clues for a new approach to resolve this issue.

2.3.2 Why Does Plate Tectonics Occur Only on Earth and How Does It Affect the Habitable Environment?

Plate tectonics is a unique feature of the Earth, which is quite different from the other terrestrial planets and is a principal factor in the development of Earth habitability. The role and function of tectonic activity in the construction of the habitable environment are key questions in the study of the habitable Earth. Study of the emergence and evolution of plate tectonics should be placed in this context.

The timing of the beginning of plate tectonics is a contentious issue. Previous studies primarily examined clues found in ancient rocks to identify the beginnings of subduction. Whereas, new perspectives and methods are required to establish a definitive timeline. In fact, plate tectonics occurred as a result of the transition from pipe-like structures created by vertical movements into more complex structures associated with horizontal movements. This transition should be the focus of future study, particularly to determine the driving force of subduction zones and the reasons for variations in thermal system through time. This will be helpful in accurately determining the initiation, characteristics, and evolution of plate tectonics in early Earth.

Specific questions include: What are the sources of driving force for the global-scale plate tectonics? What are the mechanisms? When and how did plate tectonics initiate? How are plate tectonics maintained? How do subduction slabs interact with the mantle? What is the influence of tectonic activity on the surface environment?

2.3.3 Influence of Significant Geological Events on the Environment and the Evolution of Life

Major geological events have occurred throughout Earth's history. They have directly affected the development of Earth habitability and have determined the origins and radiation of the different biomes, as well as global ecological crises and mass extinctions. For example, two major oxidation events occurred in the Early and Late Proterozoic. The first led to a rapid flourishing of eucaryotes and the second to the emergence of the first complex lifeforms. Rapid melting and breakup of the lithosphere in the Neoproterozoic were accompanied by volcanic activity on a vast scale which triggered a fast transformation from extreme ice room conditions (snowball Earth) to an extreme greenhouse climate. Large-scale continental magma eruptions in the Phanerozoic resulted in multiple mass extinction events. A huge meteorite impact may have caused the well-known 'dinosaur extinction' at the end of the Cretaceous. Thus, uncovering the full effects of these crucial geological events on biological and environmental evolution is clearly essential for understanding the development of Earth habitability (Fig. 2.3).

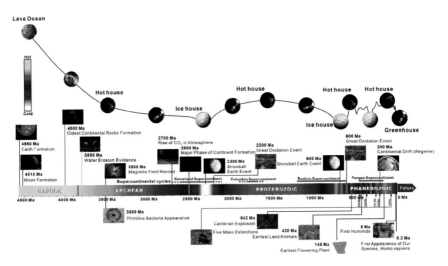

Fig. 2.3 Relationship of vital geological events to the Earth's evolving temperature cycle (drawn by Tand, Chen and Gong, according to Condie 2010; Grotzinger and Jordan 2010; Tang and Li 2016)

2.4 Earth's Atmosphere and Climate Change

2.4.1 Oceanic Heat Absorption and Carbon Sequestration, and Their Role in Climate Change

Since the Industrial Revolution, the cumulative carbon emissions due to usage of fossil fuels have reached about 450 billion tons (Friedlingstein et al. 2020), and led to a significant rise in atmospheric concentrations of CO_2. Over 90% of the heat added to the climate system (Cheng et al. 2019) and about 30% of the CO_2 emissions due to human activities (Pörtner et al. 2019) have been absorbed by the oceans. Therefore, the oceans play a crucial role in mitigating climate change. The oceans' capacities for heat absorption and carbon sequestration are determined by material and energy exchanges at the air-sea interface, between the upper and deep layers of the oceans, and at the fluid–solid interface on the sea floor (Fig. 2.4). The oceanic dynamic

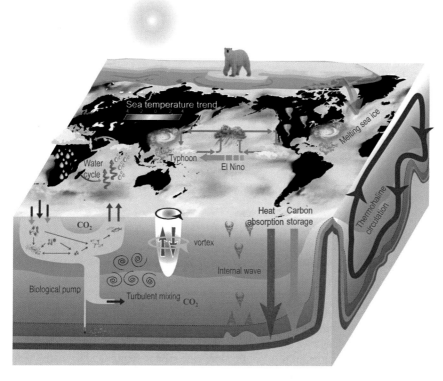

Fig. 2.4 Heat absorption and carbon sequestration processes in the ocean and their role in climate change. Ocean dynamic processes and biogeochemical processes fundamentally affect the capacity of the oceans to absorb surplus heat and human carbon emissions. Heat absorption and carbon sequestration processes in the ocean can in turn significantly alter the marine environment and exert important influences on Earth's climate system through cross-sphere interactions

processes including large-scale ocean circulation, mesoscale/sub-mesoscale eddies, and small-scale turbulence, and multi-scale air-sea interactions are the key processes for oceanic materials and energy cycle and storage. However, what is the upper limit of ocean's capacities for heat absorption and carbon sequestration in response to climate change, and where are the tipping points of climate change? These are major fundamental scientific questions to understanding the oceans' role in climate change. To address this question, it is needed to understand the role of mesoscale and smaller-scale processes, which contain the most part of ocean's energy, in heat absorption and carbon sequestration; accurately quantify the heat and carbon exchanges at the air-sea interface; and clarify the contributions of ocean's biological and physical carbon pump to absorbing atmospheric CO_2.

The continuous absorption of heat and carbon by the ocean changes the marine environment significantly, exerts enormous and even disastrous effects on Earth's climate system through cross-sphere coupling, and would significantly affect the Earth's habitability in a visible future by human beings (Fig. 2.4). However, it remains unknown when and how the continuous oceanic change would shape Earth's climate in context of global warming, and how to better predict ocean's feedback to global climate change in the future. The answers to these questions are key to understanding the habitability of future Earth. In the future, it is needed to carry out systematic research in terms of: the mechanisms through which the continuous heat and carbon absorption affects ocean's dynamics and thermal structures; the processes that determine the changes of ocean heat content, polar ice and snow, and sea level; the mechanisms through which the air-ice-ocean interactions affect global and regional climate change; and improved prediction of the frequency and intensity of future extreme weather and climate events; and a better understanding of the ocean's feedbacks to global climate change and underlying mechanisms.

2.4.2 Climate Change in Extremely Hot Geological Periods

As the largest active carbon reservoir on the Earth's surface, the ocean is the most crucial buffer against climate change in Earth's surface systems. It's fundamental in maintaining a habitable climate and a tenable ecologic environment for human life. In the geological past, the Earth has experienced extreme climate and environment change events on a vastly greater scale than those that have occurred during human history. During those periods, the atmospheric CO_2 concentration has been much higher in the past than it is now. Climate change mechanisms in hyperthermal periods of geological past and the corresponding response of oceanic ecological systems have critical reference implications for comprehending the requirements for human survival in the current conditions of accelerating environmental change driven by global warming. However, the detailed evolutionary processes and forcing mechanisms of paleoceanography, paleoclimate, and palaeontology in periods of extremely rapid warming in geological history such as, the Cretaceous anoxic events, the Paleocene-Eocene Thermal Maximum (PETM), the Eocene and Middle Miocene

Climatic Optimum, remain unclear. The opening and closure of major ocean gateways may also have played a significant role in past climatic and ecological change, but are not well understood. The following questions must be urgently addressed: What are the thresholds of abrupt changes in global climate, environment and ecosystem in response to tectonic activity and rapid emission of greenhouse gases? How did the marine ecosystem and climatic condition rapidly restore after the disaster? Are the current conditions of carbon emissions and the present rate of global temperature increase sufficient to drive a transition to a different state of the oceans' ecological system?

2.5 Interactions Between the Oceans and the Earth's Interior

2.5.1 Lithosphere Structure and Composition of Oceanic Plates

The composition and structure of the oceanic lithosphere are the keys to unlocking the driving forces of plate tectonics. In 1957, the "Moho" project—an ambitious plan to penetrate the Mohorovicic discontinuity in the sub-oceanic crust—was set out by the eminent American scholar of physical oceanology, Walter Munk, and the pioneer of plate tectonics theory, Harry Hess, amongst others (Hess and Ladd 1966). Their intention was to observe and study the lithospheric structure and composition of the sub-ocean plates at first hand by drilling through to the Earth's mantle. At the time, this plan was thought to rival the Apollo program in scale and importance. Whereas, the primary objective of the Apollo program—to send men to the moon and return them safely to Earth—was achieved half a century ago, together with the collection of large quantities of geological samples from the lunar surface, representing a giant leap in our knowledge and understanding of the moon. Nevertheless, although deep-sea drilling has been conducted throughout fifty years, the Moho discontinuity remains untouched. In recent years, an increasing number of studies have confirmed that the structure and composition of the oceanic lithosphere are very complex and quite different from classical models (Wilson et al. 2006; Gillis et al. 2014; Sutherland et al. 2017). To reach new understandings, multi-disciplinary cooperation is required, involving marine geology, geophysics, high pressure and high temperature experimental simulation, and computer science. Major questions include: How does the oceanic lithosphere thicken from mid-ocean ridges to subduction zones? What is the composition of the thickened lithospheric mantle? What are the alteration processes of the oceanic lithosphere and their controlling factors? How influential is the alteration of the oceanic crust on deep-water circulation and fluid–solid interactions?

2.5.2 Driving Forces of Oceanic Plate Movements

Plate tectonic theory is the basis of solid geoscience (Fig. 2.5). However, the nature of the forces that drive plate movements are still unclear and are hotly disputed (Zheng and Zhao 2020). Current mainstream opinions relate the driving forces of plate movement to mantle convection, mantle plume upwelling, slab pull, and the 'magma engine' hypothesis. Whereas, the rate of mantle convection is much slower than the movement of the oceanic plates, so mantle convection cannot be the principal driving force for tectonic activity (Anderson 1998). Super-mantle plume upwelling could trigger continental disintegration but could not have driven the gradual and continuous sea floor spreading and oceanic basin formation that followed the breakup of the supercontinents. Subduction slab dragging could certainly influence local plate movements but could not be responsible for large-scale ocean-floor spreading, which is not related to subduction zones, or provide sufficient force for oceanic plate re-orientation or initiation of subduction (Sun 2019; Stern and Gerya 2018; Arcay et al. 2020). The 'magma engine' hypothesis proposes that heat from the Earth's interior is the main energy source for plate movement. Newly formed oceanic crust around mid-ocean ridges is light and thin, while old oceanic crust is heavy and thick. This combination causes tectonic plates to be inclined relative to the asthenosphere, generating sliding forces that cause continuous mid-ocean ridge spreading driven by sinking of old crust and the formation of more new oceanic crust by upwelling of magma. The sinking of old, heavy crust in subduction zones therefore provides the power to drive this vast 'engine' (Sun 2019). However, this hypothesis has never been verified. Questions to be resolved include: Where does the energy driving plate movements come from? Why do plate movements occur only on Earth of all the terrestrial planets? What role does fluid–solid interaction led by sea water play in

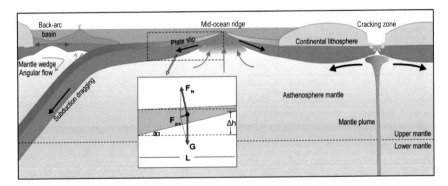

Fig. 2.5 Sketch of the driving forces of plate movement. Verification defects occur in the driving force models of slab dragging, mantle convection and plumes. The rate of mantle convection is slow. In addition, the flow direction of the mantle at mid-ocean ridges is opposite to the direction of plate movement. Plate movement without subduction cannot be caused by subduction slab dragging. Continuous plate movement cannot be driven by mantle plumes. The magma engine hypothesis, although promising, requires further verification

plate movements? How can we verify the 'magma engine' driving force hypothesis for plate movements?

2.5.3   Deep-Water Circulation and Sea Level Fluctuation

At present, about 60% of the world's population reside in coastal areas, i.e., within 100 km of the sea, so that sea level fluctuations have a huge influence on the human habitable environment. Traditional studies of sea level fluctuations have focused on ice cap melting, glacio-isostatic adjustment, variations in sea water density, changes in inland water storage conditions, vertical displacement caused by movements of the crust, and the overall influence of sea water motion. The amount of water in the Earth's interior exceeds that held in the surface oceans (Langmuir and Broecker 2012), with plate subduction and magmatic activity controlling water circulation in the deep Earth (Cai et al. 2018; Galvez et al. 2016; Parai and Mukhopadhyay 2018). With respect to Earth's evolution, sea level fluctuation is defined by the total amount of sea water and the average age of the oceanic crust, with the former being primarily controlled by water exchange between the Earth's interior and the surface driven by plate subduction and hydrothermal fluid processes in the mantle, with the latter controlled by super-continent convergence and breakup and super-ocean opening and closing. These geological processes have also controlled ice cap formation and dissolution in the past. For example, during the event known as the Permian Extinction, volcanic activity in large igneous provinces, combined with plate subduction, released vast quantities of greenhouse gases in a concentrated discharge over a relatively short period of time, causing rapidly increasing air temperature and creating a greenhouse climate with no polar ice caps. The result was the extinction of almost 80% species on Earth (Fig. 2.6). Key issues in this area include: What is the relationship between the exchange flux of water in the Earth's interior and sea water on the surface? How does the total volume of surface water vary? Will Earth's oceans dry up like those on Mars and when might that happen? How do "greenhouse" and "ice house" conditions on Earth flip from one state to the other? How do large-scale geological events affect the formation and melting of the ice caps?

2.6   Evolution of the Oceans and the Origins of Life

2.6.1   Extreme Oceanic Life Processes and Origins of Life

We know that life originated on Earth no later than 3.8 Ga, but the mechanisms that created the conditions for life remain highly controversial. In 1953, Harold Urey and Stanley Miller synthesized amino acids from reductive gases (ammonia and methane, etc.) using electrical discharges. Therefore, intense lightning storms in the strongly

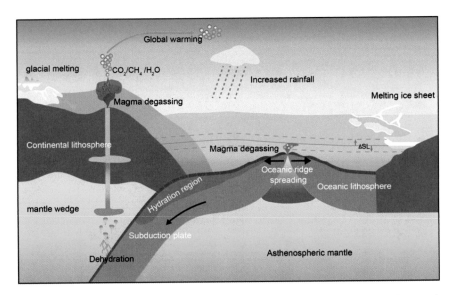

Fig. 2.6 Sketch of deep-water circulation and sea level fluctuation. Sea water enters the oceanic lithosphere by seafloor seepage and is carried into Earth's interior by plate subduction. It is then released to the surface by magmatic and hydrothermal activities, completing the deep-water circulation cycle. On a geological scale, this process controls sea level fluctuations

reductive primordial atmosphere have been considered as the key factor in creating the chemical precursors of life on Earth. Nevertheless, subsequent studies have shown that the atmosphere of early Earth was not as reductive as originally thought, casting doubt on the mechanism suggested by the Urey-Miller experiment. The current mainstream view is that localized extreme environment (deep-sea hydrothermal fluids and cold springs, etc.) were probably the original sites of life (Fig. 2.7). For instance, alkaline thermal fluids associated with serpentinization at low temperature generate large amounts of methane, which could support chemosynthetic ecosystems and have been the hotspot for research of the origins of life. The study of lifeforms in extreme deep-sea environment could assist us in understanding the origins of life and the coherent evolution of organisms and the environment. Scientific issues to focus on in the future include: key chemical and physical processes in the origin of life; deep-sea serpentinization and transition mechanism from inorganic to organic; the coupling mode and mechanism between the origin of life and major geological processes; hypothesis of origin and evolution of photosynthetic organisms in deep-sea hydrothermal fluids; the deep-sea origin hypothesis of eukaryotic life.

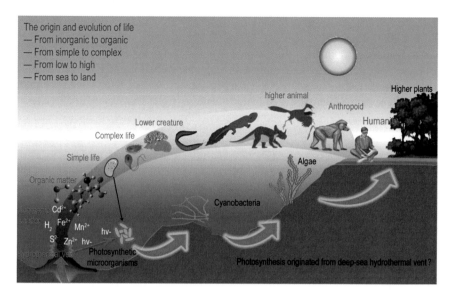

Fig. 2.7 Life origin and evolution hypothesis

2.6.2 Evolution of Marine Life and Its Adaptability to the Environment

The oceanic environment is complex and variable, so marine organisms have evolved a wide range of adaptation mechanisms. Such organisms can survive independently in hydrothermal fluids, cold springs, abysses and other extreme environment, and have also evolved a high level of biological diversity. Studies of deep marine life indicate that physical and chemical factors, as well as geological processes, shape the environmental constraints on marine life (Heuer et al. 2020). However, the evolutionary processes and mechanisms under environmental change are still unclear. Important issues awaiting clarification and interpretation include: the limits and controlling factors of life in extreme marine environment; epigenetic and genetic mechanisms of adaptive evolution during the migration of deep-sea organisms between habitats; the driving processes and mechanisms of diversity differentiation in marine organisms from different environment; variation, response strategies and ecological function changes of marine organisms in the context of global change.

2.6.3 Role of Marine Life in Earth Evolution

Marine life has had a significant effect on Earth's evolution by its influence and participation in substance circulation and energy flow. Organic matter from the ocean surface was gradually degraded by marine organisms and was deposited in

water bodies and sediments, entering into material circulation in the lithosphere and
Earth's interior through diagenesis (Fig. 2.8), controlling the formation of hydrocarbon resources and marine gas hydrates. Against the global background of climatic
variation, the community structures and ecological functions of marine organisms
have changed markedly, and their roles in generating feedback phenomena in climate
change are now receiving unprecedented attention (Cavicchioli et al. 2019). Crucial
questions that call for breakthroughs in this field include: What are the associations
between deep-sea biological activity and specific geological structures and events?
What are the contributions and ecological effects of nutrient transportation by marine
organisms to surrounding ecological systems? How are materials transformation and
energy flow driven by marine organisms and what are their roles in regulating Earth's
climate system (for example, the mechanisms of methane leakage, their scope, and
their influence on global warming; the contribution of biogenetic sulfur to cloud
formation and negative greenhouse effects; etc.)?

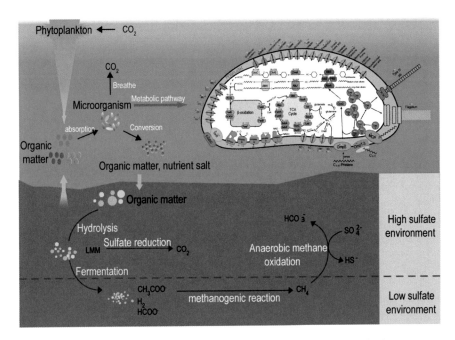

Fig. 2.8 Organic matter decomposition and transformation driven by microbes in deep-sea water
bodies and sediments. After hydrolysis and fermentation, organic matter precipitates and begins to
generate methane. Most of the methane is consumed by anaerobic oxidization before being released
into water bodies

2.7 Multi-sphere and Multi-scale Interaction Processes and Physical Mechanisms of the Sun and Earth in a Deep-Space Environment

2.7.1 Sun and Earth Interaction

Solar activity can vastly alter the energy and state of Earth's atmosphere and ionosphere and can induce spatial weather variations, leading to changes in the lower atmosphere and even on Earth's surface. Solar activity is therefore a non-negligible driving element for global change and for terrestrial environmental catastrophes (Bothmer and Daglis 2007; Kamide and Chian 2007; Zhang et al. 2012). Research has established that multi-scale space weather effects are closely related to Earth habitability and have a significant nonlinear influence on aspects of the evolution, transmission, and integration of Earth's multi-sphere systems. A shortage of empirical observations and numerical simulations means that many important questions relating to mutual sphere boundaries and dynamic structural boundaries are still to be resolved on both macro- and microscopic scales. With respect to perception of the multi-sphere activity of the sun and the Earth and their influence on terrestrial geological activity, essential researches will involve: determining the mechanisms by which meteorological systems, landform variations, and lithospheric changes stimulate atmospheric fluctuations; assessing the influence of crustal activity on atmospheric ionization and electrostatic fields; and investigation of electromagnetic connections and chemical diffusion processes between the lithosphere, atmosphere, and ionosphere.

2.7.2 Geological Activity and the Formation Mechanisms of Icy Celestial Bodies

Across the solar system, clear traces of a variety of geologic processes have been identified on the surfaces of icy celestial bodies, for example, impact meteorite craters, ice volcanoes, faults, mountain ridges, and chaotic terrain areas. From our current state of knowledge, we can identify the geological processes that created many of these features. However, the geneses of other features are still unknown. For instance, normal extensional faults are fairly common on ice satellites, such as Jupiter's moon, Europa, but reverse compression faults are rare. There are large areas of chaotic terrains on the surface of Europa, but their geneses are entirely unknown at present. How do these surface structures form? What are the deep structures of the ice crust? What are the patterns and extent of internal convection beneath the ice crust? Do both superficial and deep ice circulation occur? What is the nature of the boundary between the lower surface of the ice crust and the salt-water ocean that is believed to exist beneath the frozen surface of Europa? Does subduction occur in the ice crust? What about plate tectonics?

With respect to other icy celestial bodies, essential questions are as follows: What are the geneses of ice volcanoes? What are the internal structures of these bodies? How many impact craters are there and how are they distributed? What were the origins of the meteoroids? Could internal structural features be produced by viscouse relaxation? Are there active processes and features, such as liquid methane lakes, on the surface of Titan, Saturn's largest moon? What is the composition of Pluto's surface mountains and how were they formed?

2.7.3 Study and Evaluation of Small Extraterrestrial Celestial Bodies, and Prediction and Prevention of Earth Impacts

The possibility of a small celestial body impacting the Earth presents a significant threat to human existence. In 1908, a relatively small meteor (30–50 m in diameter) exploded in the sky over Tungus, Russia, destroying over 2000 km^2 of Siberian forest. In 2013, another small object (only about 18 m in diameter) impacted the Chelyabinsk Area, Russia, (Brown et al. 2013) injuring around 1500 people and damaging over 3000 buildings. Nevertheless, actual meteorite impacts (meteoroids that strike the Earth's surface) are comparatively rare, since ablation, ariel explosions, and disintegration generally destroy small celestial bodies as they traverse the atmosphere. The atmosphere therefore represents Earth's primary defense against ground impacts. Studies of small extraterrestrial celestial bodies must determine the following: the occurrence rate of gamma-ray bursts, calculated independently of existing models based on historical observation; the evolutionary laws of celestial bodies in the solar system and the flux of small meteoroid impacts on the Earth; the filtration effect of Earth's atmosphere on small celestial bodies and the historical incidence of meteoroid events; the nature of the two-way material exchange processes between Earth and interstellar space; the likelihood of meteorites of earth origin being found on the moon and Mars; and, of course, the available mechanisms for defending the Earth against small celestial body impacts.

2.8 Detection of Habitable Extrasolar Planets

What information from planetary observations and measurements is useful in determining whether a planet is habitable or not? Based on understandings gained from observation of the planets in our own solar system, four factors need to be considered to answer this question. Firstly, the planet must be neither too small nor too large. Planets that are too small will be unable to maintain stable atmospheres or magnetic fields, while planets that are too large are likely to be gaseous, like Jupiter and Saturn in our solar system. One rough measure is that planets within 0.1–5 times the size of Earth have at least the potential to be habitable. Secondly, habitable planets must be located in stellar zones falling within the astronomical definition of habitability, which is related to the characteristics of the stars around which the planets orbit. Thirdly, detection of atmospheric components in extrasolar planets (e.g., particularly water vapor, O_2, O_3, and CH_4) is clearly helpful in assessing the likelihood of a planet harboring life—or at least life similar to Earth's. Lastly, whether seas, land, glaciers, and organic matter exist on the surface of extrasolar planets are useful pointers in considering habitability. Recently, liquid water and atmospheres (mainly consisting of hydrogen and helium) have been discovered on super-terrestrial planets. Whereas, it is still difficult to determine whether these planets have plate tectonics similar to the Earth. In the future, the fundamental question of "What are the basic conditions required for life to exist in the universe?" is one that must be answered. Our understanding of this most basic question is still evolving. It is now known, for example, that atmospheric air is not necessary for life in submarine environments, although, as far as we know, it is a universal feature of life on Earth's surface. What, in fact, are the true necessities of terrestrial life? It is a fundamental question which must be answered if we are to successfully predict and detect life in deep space. The apparently essential conditions of life on Earth's surface are not necessarily the essential conditions for life on other planets, or even in other spheres of the earth.

2.9 The Human-Earth System and Sustainable Development

Global climate change and anthropogenic activities have intricated plenty issues in environment, resources and ecology. These problems are continuously challenging the sustainable development of human society and the ecological civilization construction in our country. To resolve these problems, the critical basis is to comprehensively understand the interaction between human and Earth systems, and further to uncover the dynamic mechanisms of human-Earth systems and the mechanisms required for sustainable international, national, and regional development (Fu 2020).

The human-Earth system is considered to be a dynamic combination of human society and natural systems; thus, its structure and feature are different from either single system. The human-Earth system has many possible stable states, it allows conversion between different stable states when critical thresholds are surpassed. Understanding of the dynamic evolution of the overall human-Earth system and the feedback mechanisms between its subsystems will require study of the elasticity, adaptability, and transformative capacity of the entire system (Fig. 2.9). The human-Earth system has multiple challenging properties, including complexity, nonlinearity, uncertainty, and multi-sphere nesting. Driven by global change and human activities, both social and natural systems are now in accelerating states of dynamic variation.

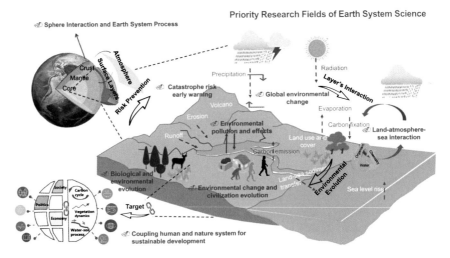

Fig. 2.9 Preferred strategic areas for the study of the development of the human-Earth system

Description of the evolutionary laws of human-Earth system structures and functions, and uncovering the dynamic mechanisms that support them, will provide the scientific basis for maintaining and enhancing the elasticity and adaptability of the system and for promoting regional sustainable development.

In the human-Earth system, the natural geographic system is the basic parameter of socio-economic dependency. The combination of the multiple elements and processes of water, soil, air and ecology determines the functionality of entire regions and their capacities to bear demands on their resources, which present as complete ecosystem services.

Long-term observation, field test, and numerical simulation studies are essential to understand the ecological structures and the relationship between the important ecological processes and its functions and services. This will enable clarification of how changes in ecologic system structures and processes affect the generation, transmission, and actualization of ecological system functions and services. Revealing the connections between ecologic system services and human well-being is the basis for comprehending the values of ecological system services, the demands that are placed on them, and their internal associations with human social well-being (Wang et al. 2013). To improve and optimize the ecosystem services, it is essential to know the linkage of the demands put on the ecosystem services and the social well-being, to distinguish the interaction mechanism between the dynamic changes of ecosystem services and human well-being and sustainability.

2.10   Global Environmental Change and the Evolution of Biology and Human Culture

Since Earth was born, the terrestrial environment has been continuously changing, driven by a combination of internal and external forces to form the fundamental environmental framework of Earth as it exists today. Since the Industrial Revolution, the influence of human activities on Earth's environment have accelerated and intensified. The global biogeochemical cycle is affected by the changes to land surface system mainly induced from human activities (e.g., surface alteration and construction, and pollutant discharge). The ultimate aim of research on global environmental change is to avoid or mitigate the destructive impact of human activities on Earth's ecology and environment while promoting the sustainable development of human society.

To achieve these vital objectives, future studies must focus on: (1) identifying the mechanisms by which human activities influence global environmental change, particularly in their effects on biogeochemical cycle between the multiple surface spheres and connecting media; (2) investigating the interactions and mutual influence of regional and global environmental change; (3) predicting the future influence of human activities on the environment and the reciprocal influence of environmental change on human society, especially for the prediction of the critical thresholds of destructive and catastrophic environmental changes; (4) developing strategies to allow human society adapting to the environment change and mitigating the impact of human activities on the global environment.

To study the co-evolution of life and environment, it requires intersection and deep integration of multiple disciplines. For example, what is the relationship of biodiversity to variations in the concentrations of CO_2 and O_2 in the atmosphere and ocean? How do deep-Earth activity and continental variations impact the evolution of Earth's ecological systems? How do extreme climate events affect the stability of Earth's ecological systems? What is the current and future influence of global warming on biodiversity in the ocean and on land? Can a sixth mass extinction be avoided? What are the differences in adaptability to environment change between biomes? How contemporary environmental change driving spatiotemporal variations in biodiversity? Current studies in biology, molecular biology, evolutionary developmental biology, and epigenetics indicate that environmental factors can both directly and indirectly influence biological phenotypes by controlling their genetic expression. Hence, besides the effects of environment as a major factor in natural selection, the interactions between genetics and the environment and their influence on biological evolution will become major research directions.

2.11   Sphere Interactions and Earth System Processes

The Earth's plate tectonics are significant pathways for materials and energy exchange among Earth's spheres and are essential to Earth's evolution as a habitable planet. The study of sphere interaction is an effective approach for understanding the past, present, and future of Earth habitability (Fig. 2.10). For instance, Earth's magnetic field, generated by a geomagnetic 'dynamo' process driven by Earth's inner and outer cores, can protect our atmosphere from erosion by the solar wind. The dynamic processes of Earth's interior maintain the oceans in which the complex submarine life systems generate and further originate life on Earth. In addition, at different evolutionary stages of the habitable Earth, the deep Earth processes and the connections between internal and external systems may have varied (Brune et al. 2017; Campbell and Allen 2008; Smit and Mezger 2017). This variation could increase the difficulty in explaining the mechanisms of interactions between interior and exterior Earth systems. To address this challenging issue, we can focus priory

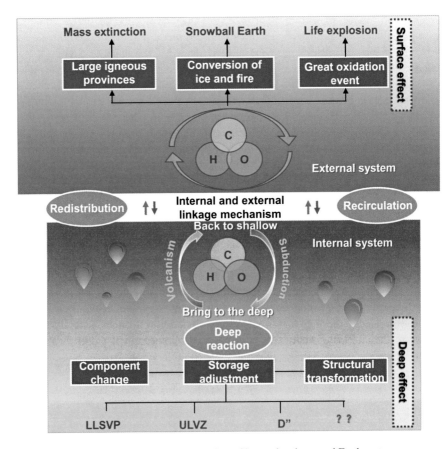

Fig. 2.10 Working model of linkage mechanism of internal and external Earth system

on the cycle behavior of the vital life elements (C–H–O–S) in the deep Earth at several major geological events. Specifically, its necessities to investigate the laws governing the cycle and re-distribution of the elements (C–H–O–S) in the deep Earth and underlying mechanisms. It is also vital to understand the exchange of materials and energy between the super-deep and upper zones of the Earth's interior as well as their roles in both gradual global evolution and the abrupt change. The synthesis of multi-disciplinary studies, big data analysis and Earth dynamic simulation will clarify the mechanisms controlling the interactions between inner and outer Earth spheres and their effects on the planet's habitability. In this way we will be able to construct theories which will reveal the deep controlling mechanisms of Earth habitability.

Materials circulation in the deep Earth may play a critical role in shaping Earth habitability, mainly through the long-term reconstructive effects of volcanic activity, plate tectonic activity, and fluid circulation in surface systems (Sleep and Zahnle 2001; Foley and Fischer 2017). Crust-scale fluid activity occurs widely in inner ocean plate before it enters subduction zones (Fig. 2.11). Oceanic subduction zones are sites for carbon exchange between the exterior and interior of the Earth. Some carbon enters the mantle, while the remainder is returned to the surface along with island arc magma (Kelemen and Manning, 2015). This process raises a number of questions: Does plate subduction contribute to the generation of convective driving forces in the deep mantle? How do carbon and water circulate in the deep Earth and on the surface? To get innovative thoughts for us to recognize the sphere interactions

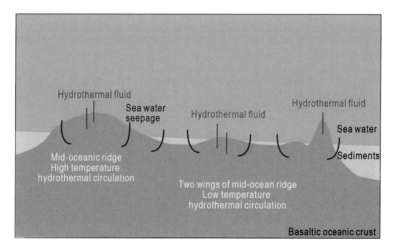

Fig. 2.11 Schematic of submarine hydrothermal fluid circulation. Submarine hydrothermal fluids and cold springs are important channels for materials and energy exchange between the oceans and the oceanic crust. Hydrothermal activity mainly occurs in mid-ocean ridges, with extensive mineralization and development of hydrothermal ecological systems. In areas away from mid-ocean ridges, methane leakage in sediments is the principal factor in the formation of cold springs. The hydrogen and methane generated in serpentinization were the primary cold springs in Earth's early history

in Earth's systems, we need to consider the interactions between planets and the interactions of multiple spheres on Earth, to explore the fundamental laws governing the periodic fluctuations of cold and warm conditions on Earth, to investigate major geologic events and their internal systemic relationships with thermal fluctuations in Earth's evolution and to quantify the relative contributions of natural and human activities on global climate change.

References

Anderson DL (1998) The scale of mantle convection. Tectonophysics 284(1–2):1–17

Arcay D, Lallemand S, Abecassis S et al (2020) Can subduction initiation at a transform fault be spontaneous? Solid Earth 11(1):37–62

Armstrong K, Frost DJ, Mccammon CA et al (2019) Deep magma ocean formation set the oxidation state of Earth's mantle. Science 365(6456):903–906

Bothmer V, Daglis IA (2007) Space weather—physics and effects. Praxis Publishing, Chichester, p 487

Brown PG, Assink JD, Astiz L et al (2013) A 500-kiloton airburst over Chelyabinsk and an enhanced hazard from small impactors. Nature 503:238–241

Brune S, Williams SE, Müller RD (2017) Potential links between continental rifting, CO_2 degassing and climate change through time. Nat Geosci 10:941–946

Cai C, Wiens DA, Shen WS et al (2018) Water input into the Mariana subduction zone estimated from ocean-bottom seismic data. Nature 563:389–392. https://doi.org/10.1038/s41586-018-0655-4

Campbell IH, Allen CM (2008) Formation of supercontinents linked to increases in atmospheric oxygen. Nat Geosci 1(8):554–558

Cavicchioli R, Ripple WJ, Timmis KN et al (2019) Scientists' warning to humanity: microorganisms and climate change. Nat Rev Microbiol 17:569–586

Cheng L, Ibraham J, Hausfather Z et al (2019) How fast are the oceans warming? Science 363(6423):128–129

Condie KC (2010) Earth as an evolving planetary system. Elsevier, Amsterdam

Dasgupta R (2013) Ingassing, storage, and outgassing of terrestrial carbon through geologic time. Rev Mineral Geochem 75(1):183–229

Dessert C, Dupré B, Gaillardet J et al (2003) Basalt weathering laws and the impact of basalt weathering on the global carbon cycle. Chem Geol 202(3–4):257–273

Foley SF, Fischer TP (2017) An essential role for continental rifts and lithosphere in the deep carbon cycle. Nat Geosci 10:897–902

Friedlingstein P, O'Sullivan M, Jones MW et al (2020) Global carbon budget 2020

Fu BJ (2020) Sustainable development Goals of The United Nations and the historical mission of geographic science (in Chinese). Sci Technol Rev 38(13):19–24. https://doi.org/10.3981/j.issn.1000-7857.2020.13.002

Galvez ME, Connolly JAD, Manning CE (2016) Implications for metal and volatile cycles from the pH of subduction zone fluids. Nature 539:420–424

Gillis KM, Snow JE, Klaus A et al (2014) Primitive layered gabbros from fast- spreading lower oceanic crust. Nature 505:204–207

Grotzinger J, Jordan T (2010) Understanding Earth, 6th edn. W.H. Freeman and Company, New York, p 672

Hess HH, Ladd HS (1966) MOHOLE: Preliminary Drilling. Science 152:544–545

Heuer VB, Inagaki F, Morono Y et al (2020) Temperature limits to deep subseafloor life in the Nankai Trough subduction zone. Science 370(6521):1230–1234

Hou MQ, Zhang Q, Tao RB et al (2019) Temperature-induced amorphization in $CaCO_3$ at high pressure and implications for recycled $CaCO_3$ in subduction zones. Nat Commun 10(1):1963

Hu QY, Kim DY, Liu J et al (2017) Dehydrogenation of goethite in Earth's deep lower mantle. Proc Natl Acad Sci USA 114(7):1498

Hu QY, Liu J, Chen JH et al (2020) Mineralogy of the deep lower mantle in the presence of H_2O. Natl Sci Rev. https://doi.org/10.1093/nsr/nwaa098

Kamide Y, Chian A (2007) Handbook of the solar-terrestrial environment. Springer, Berlin, New York, pp 539

Kelemen PB, Manning CE (2015) Reevaluating carbon fluxes in subduction zones, what goes down, mostly comes up. Proc Natl Acad Sci USA 112(30):3997–4006

Langmuir CH, Broecker W (2012) How to build a habitable planet: the story of Earth from the big bang to humankind, Revised and Expanded. Princeton University Press, Princeton, p 736

Liu J, Hu QY, Kim DY et al (2017) Hydrogen-bearing iron peroxide and the origin of ultralow-velocity zones. Nature 551(7681):494–497

Liu JW, Zheng YF, Lin HY et al (2019a) Proliferation of hydrocarbon-degrading microbes at the bottom of the Mariana trench. Microbiome 7:47

Liu QY, Zhu DY, Meng QQ et al (2019b) The scientific connotation of oil and gas formations under deep fluids and organic-inorganic interaction (in Chinese). Sci China Earth Sci 62:507–528

Mao HK, Mao WL (2020) Key problems of the four-dimensional Earth system. Matt Radiat Ext 5:038102

Mao HK, Hu QY, Yang LX et al (2017) When water meets iron at Earth's core-mantle boundary. Natl Sci Rev 4(06):870–878

Mcnamara AK (2019) A review of large low shear velocity provinces and ultra-low velocity zones. Tectonophysics 760:199–220

Mundl-Petermeier A, Walker RJ, Fischer RA et al (2020) Anomalous ^{182}W in high^{3}He/^{4}He ocean island basalts: fingerprints of Earth's core? Geochim Cosmochim Acta 271:194–211

Olson P, Landeau M, Reynolds E (2018) Outer core stratification from the high latitude structure of the geomagnetic field. Front Earth Sci 6:140

Parai R, Mukhopadhyay S (2018) Xenon isotopic constraints on the history of volatile recycling into the mantle. Nature 560(7717):223–227

Pörtner HO, Roberts DC, Masson-Delmotte V et al (2019) The ocean and cryosphere in a changing climate. In: A special report of the intergovernmental panel on climate change. IPCC, pp 765

Schmidt MW, Poli S (2014) Devolatilization during subduction. Treat Geochem 4:669–701

Sleep NH, Zahnle K (2001) Carbon dioxide cycling and implications for climate on ancient Earth. J Geophys Res Atmos 106(E1):1373–1399

Smit MA, Mezger K (2017) Earth's early O_2 cycle suppressed by primitive continents. Nat Geosci 10:788–792

Stern RJ, Gerya T (2018) Subduction initiation in nature and models: a review. Tectonophysics 764:173–198

Sun WD (2019) The Magma engine and the driving force of plate tectonics (in Chinese). Chin Sci Bull 64:2998–3006. https://doi.org/10.1360/N972019-00274

Sutherland R, Townend J, Toy VG et al (2017) Extreme hydrothermal conditions at an active plate-bounding fault. Nature 546(7656):137–140

Tang CA, Li SZ (2016) The Earth evolution as a thermal system. Geol J 51(S1):652–668

Wang S, Fu BJ, Wei YP et al (2013) Ecosystem services management: an integrated approach. Curr Opin Environ Sustain 5(1):11–15

White WM (2015) Isotopes, DUPAL, LLSVPs, and Anekantavada. Chem Geol 419:10–28

Wilson DS, Teagle DAH, Alt JC et al (2006) Drilling to gabbro in intact ocean crust. Science 312:1016–1020

Zhang L, Wang C, Fu SY (2012) Solar variation and global climate change. Chin J Space Sci 31(5):549–566

Zhang X, Li LF, Du ZF et al (2020) Discovery of supercritical carbon dioxide in a hydrothermal system. Sci Bull 65(11):958–964

Zheng YF, Zhao G (2020) Two styles of plate tectonics in Earth's history. Sci Bull 65(4):329–334

Chapter 3
Basic Scientific Issues Relating to Earth Habitability

3.1 Important Issues in Resources and Energy Security

3.1.1 Oil and Gas Resources

Oil and gas are strategically essential for China's national energy security and economic development. In 2019, the total global primary energy consumption was 14.05 billion tons of oil equivalent. Among them, the consumption of crude oil, natural gas, coal and non-fossil energy sources accounted for 33.1%, 24.2%, 27%, and 15% of the total, respectively (BP 2020). It is estimated that oil and natural gas will still provide about 50% of global primary energy in 2050, which means that fossil fuels will continue to dominate the primary energy consumption for the foreseeable future.

Current theoretical studies in oil and gas geology integrate the interactions between Earth's spheres with their effects on resources and the environment. Since the Yanshan Movement in Eastern China, a series of petroliferous basins have formed in Northeast Asia under the influence of the deep dynamic process of the continuous subduction of the Western Pacific Plate into the Eurasian Plate (Li et al. 2010; Zhu et al. 2015; Meng 2017). These eastern petroliferous basins have unique geographical advantages and great scientific research value which can offer scientists a natural laboratory for exploring deep-surface interactions, organic–inorganic interactions, and land-sea interactions, material circulation, energy migration, etc. Geologists from around the world have documented a large number of research results and innovative understandings about the geological impact of Western Pacific Plate subduction on Northeast Asia (Sun et al. 2008; Jin et al. 2007; Zhu and Xu 2019). However, the precise coupling mechanism between the subduction process of the western Pacific plate and the oil and gas resources in the eastern basins of China has not been established. Researchers should pay attention to the key mechanisms closely related to the study of petroleum geology, not only to the interactions between the Earth's

© Science Press 2022, corrected publication 2022
The Research Group on Development Strategy of Earth Science in China,
Past, Present and Future of a Habitable Earth, SpringerBriefs in Earth System Sciences,
https://doi.org/10.1007/978-981-19-2783-6_3

spheres and organic–inorganic interactions, but also to the deep-Earth processes and thermal dynamics, marine processes and the terrestrial environment, Earth system processes, and the oil and gas generation and supply potential (Zhang et al. 2017; Liu et al. 2019a, b). Three issues are crucial to study of the coupling mechanisms linking subduction of the Western Pacific Plate with generation of oil and gas in eastern basins of China.

(1) Understanding the processes of sedimentation-diagenesis-transformation, the mechanisms of source-reservoir-cap development, and the laws governing distribution of petroliferous basins undergoing dynamic transformation by plate subduction. The formation of petroliferous basins in eastern China is controlled by plate tectonics, lithospheric structure, and the crust-mantle and lithosphere-asthenosphere interactions. A crucial frontier scientific issue in petroleum geology is to study the source-reservoir-cap development mechanisms, dynamic evolution, and distribution laws of the petroliferous basins and their correlations with plate movements and mantle convection.

(2) Determining the interactions of the Earth's spheres, the dynamics of material and energy transmission and conversion under high temperature and pressure conditions, and the mechanisms of oil and gas generation, migration, accumulation, and preservation. Oil/gas generation, migration, accumulation, and preservation is a thermal/dynamic process with circle interaction. Oil and gas are formed in particular tectonic environments—accumulation of material and energy of the Earth's lithosphere, hydrosphere, biosphere, and atmosphere under the physical, chemical, and biological actions during Earth's evolution.

(3) Organic–inorganic interactions have been ubiquitous throughout the Earth's evolution. A better understanding of the reservoir accumulation and preservation mechanisms and the distribution laws of various substances (including hydrocarbon fluids) in the deep Earth, such as solids, liquids, and gases, will help to reveal the dynamic mechanisms of deep fluid migration into basins, the controlling effect of deep geological structures on hydrogen-rich and hydrocarbon-rich deep fluids, the material and energy exchange pathways of deep fluids in basins, etc.

The research on the coupling mechanism between the subduction of the western Pacific plate and the oil and gas resources in the eastern basins of China can not only provide theoretical support for the increase of oil and gas reserves and production in the eastern basins of China, but also promote international cooperation in the study of active continental margin petroliferous basins around the world. The large-scale volcanic activity and source rock development are the focus in this research. High productivity of organic matter and favorable conditions for its preservation are fundamental to the formation of high-quality source rocks. Large-scale volcanic activity is a potential geological agent driving global climate change and biological extinctions, which greatly affect the enrichment and preservation of organic matter. Volcanic materials produced by terrestrial or submarine eruptions migrate into lakes, oceans and other water bodies. Inorganic elements and inorganic salts produced by hydrolysis promote the proliferation or extermination of organisms in water bodies, impacting the development of high-quality source rocks. Volcanic ash and sulfide-containing gases dissolve in water, creating a reducing aquatic environment. Ferruginization and sulfurization protect organic matter to a certain extent. Previous studies have focused on the impact of volcanism on single factors such as paleo-productivity and sedimentary environment. However, large-scale volcanic activity not only influences the Earth's atmosphere, hydrosphere, biosphere, and lithosphere, but also transports materials up from the deep mantle that alter the original ecosystem and biological community, thereby affecting organic-rich paleo-productivity and preservation conditions. The impact of volcanic activity on organic-rich source rocks should therefore be comprehensively analyzed from the perspectives of paleo-climate, paleo-sedimentary environment, hydrocarbon generating organisms, and early diagenetic evolution. A coupling mechanism of volcanic activity and organic-rich source rocks under the comprehensive action of multiple layers should be established. The core scientific issues mainly include: (1) the spatial relationships and geochemical responses between large-scale volcanic activity and organic-rich source rock accumulation; (2) the influence of volcanic activity on paleo-climate and paleo-sedimentary environment and its controlling effect on hydrocarbon generating biological assemblages of organic matter; (3) the coupling mechanism between large-scale volcanic activity and the development of the organic-rich source rocks.

3.1.2 Mineral Resources

Generally, the formation of mineral deposits in shallow Earth is directly or indirectly related to crust-mantle structure, substances cycle, and energy transmission. Most endogenetic deposits in shallow and surface regions were originally formed in the deep Earth and subsequently reached the superficial position as a result of uplifting and denudation. High-conductivity and low-velocity blocks in the upper mantle correspond well to the concentrated deposits in the upper crust. Low-velocity blocks may have been produced by metasomatic re-enrichment in the lithospheric mantle. Therefore, the study of metallogenic mechanisms, especially study of core scientific issues such as the enrichment of minerals, must be closely integrated with systematic study of deep geological processes. However, the current research on metallogenic mechanisms mostly focuses on mineralization processes, and metallogenic fluid and mineral precipitation, while the controlling effects of deep geological processes on the migration and enrichment of ore-forming materials are still poorly understood. The precise relationship between the crust-mantle structure and large ore concentration areas has not been revealed. In view of this, in the future study of mineral resources, the following aspects should be focused on: (1) The geochemical behavior and metallogenic specificity of key metal elements; (2) The precise definition of the contribution of oceanic crust subduction and collision process to ore-forming sources and ore-forming properties; (3) The macro-controlling mechanism of crust-mantle structure and mantle metasomatism on large ore concentrations; (4)

The effect of the global deep-Earth activities and major events in specific geological periods on the large-scale supernormal metal enrichment.

The Central Asian metallogenic domain and the Tethyan and circum-Pacific metallogenic domain have both undergone complex dynamic processes such as the creation of multi-block collages, mutual superposition, and subduction/collision. The Central Asian metallogenic domain was formed by multiple rounds of convergence and aggregation of the ancient continental crust and new geological bodies. It is a gathering area of two types of geological bodies, including the closure of the Paleo-Asian Ocean and the new-born crust and ancient land blocks. It is a geological body gathering area formed through mountain bend structures, large-scale rotations, and multiple accretion. The accumulation areas dominated by island arcs include the large-scale arc-basin system in south Siberia and the Junggar, and the young crust in the east. Accumulation areas dominated by continental blocks include the areas around Balkash and Ili (Xiao et al. 2019). The Central Asian metallogenic domain was later superimposed and transformed by multi-continental collisions and the remote effects of the southern Tethyan metallogenic domain, forming a more complex pattern of plate accretion, collision, and transformation (Xue et al. 2020; Li et al. 2019). The period of frequent alternating collisions and disintegration of ancient continental crust and formation of new geological bodies which followed was the most intense phase of magmatic hydrothermal activity. Did this phase control large-scale mineralization at its peak? What was the control mechanism? Did magmatic hydrothermal activity in the geological body accumulation area dominated by newly-formed island arc determines the types and distribution laws of accretionary deposits? Did magmatic hydrothermal activity in the geological body accumulation area dominated by land masses control the type and distribution of deposits formed by collision and transformation (Xiao et al. 2019; Li et al. 2019)? These are the main scientific issues that urgently need to be studied in relation to the multi-phase oceanic basin subtraction and step-and-repeat processes related to the law of accretion, collision, and transformation of the mineralization in paleo-Asian tectonic domain.

3.1.3 Surface Water Resources

Since the mid-twentieth century, with rapid economic and social development and increasing population, water resources have become an increasingly prominent problem that has directly affected the social development and the human living environment. The World Water Development Report 2019 estimates that the global water consumption will increase by 20–30% from its present level by 2050 (UNESCO 2019). The average precipitation in China is about 648 mm, which is 20% lower than the global land average. The per capita water resources are about 2100 m^3, which is only 1/4 to 1/3 of the world average. In the context of global climate change, some areas have already suffered from frequent droughts, while others suffered from unprecedented flooding. The frequency of the extreme weather events, such as sudden heavy precipitation and typhoons is increasing, which has a significant impact on global water resource patterns (Future Earth Transition Team 2012; Stocker et al. 2014). In order to meet the national demand for a safe water supply against this background of climate change and human activity, two major scientific issues need to be addressed:

(1) Clarifying the laws governing the Earth's major climate patterns, predicting future changes in water resources, and revealing the evolution trends and active mechanisms of water systems under the influence of climate change and human activity. Major engineering decisions should be made based on a full understanding of likely future trends in water resources. Projects that are likely to have a serious impact on water resources should be subject to scientific regulation and control.

(2) Exploring the mechanisms of water resources supporting capacity, scientifically judging the changes in water resources capacity caused by major national strategies and global change, studying the scientific issues such as the improvement mechanism and threshold of water resources support capacity in response to environmental changes, and evaluating whether the existing water resources can support national strategies into the future.

3.2 Deep-Sea Resources, Energy Potential, and Marine Security

3.2.1 Deep-Sea Energy

For the deep-sea areas that are enriched in renewable and easily exploitable energy resources, the local effects of the output and transformation of these regional energy (such as offshore wind and tidal energy) from the ocean interior on marine dynamic processes need to be understood urgently. In addition, extensive use of the deep-sea energy has already impacted materials and energy movements in the oceans and has also caused changes in the marine environment—even marine disasters. How to make rational use of deep-sea energy is a vital scientific issue.

3.2.2 Deep-Sea Mineral Resources

Deep-sea minerals include polymetallic nodules, cobalt-rich crusts, polymetallic sulfides, and REE-rich sediments, which are characterized mostly by multiple metals in composition (Hein et al. 2013) and by nanoscale sizes of mineral particles (Hochella 2008). Therefore, the chemical compositions and material properties (Sun et al. 2003) of deep-sea minerals will be utilized. The essential scientific issues concerning deep-sea minerals to be addressed currently are inter alia identification of existing phases of associated valuable elements for comprehensive utilization, and

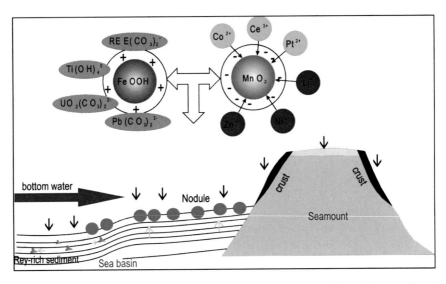

Fig. 3.1 Metallogenic model for deep-sea sedimentary deposits polymetallic nodules were formed in basins, and manganese crusts were formed on seamounts (modified from Hein et al. 2020).

clarification of material basic features. The critical scientific issues on exploration and mineralization research of polymetallic nodules, Co-rich crusts and REE-rich sediments include spatial controls on deep-sea mineral distribution, and anomalous concentration of deep-sea ore-forming elements. Essentially, those issues concern the controls of material and energy transportation in multiple spatial scales (in nanometers, centimeters, meters, and kilometers) (Fig. 3.1), namely, multiple spatial scale dynamics of deep-sea mineral sedimentation. To clarify those issues, we need to research controls from major geological and palaeoceanographical events, such as Antarctic Bottom Water, on the spatial and temporal distribution metal deposits, and controls from fluids features and ocean conditions on transportation-aggregation of metal elements in deep sea.

3.2.3 Deep-Sea Biological Resources

The deep sea harbors special microbes and large marine organisms such as fish, shrimp, crabs and shellfish, hosting many new, unknown functional genes and other active substances with specialized structures and functions (Daletos et al. 2018; Tortorella et al. 2018). In view of the abundant deep-sea biological resources, it is urgent to strengthen China's ability to excavate and utilize deep-sea biological resources (Fig. 3.2). However, the vital problem that is yet to be solved is how to obtain such resources efficiently. How to excavate microbial resources in the deep sea which are difficult to be cultured? How can we reveal the functions of the unknown

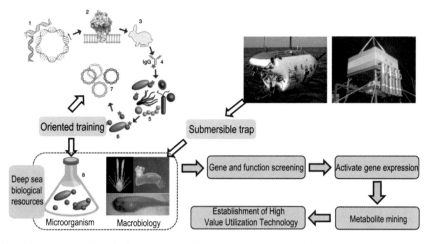

Fig. 3.2 Deep-sea biological resources acquisition and resource mining

genes in deep-sea organisms? How can we obtain active deep-sea substances that may be applied potentially in the fields of disease prevention and control, environmental protection, and in industry and agriculture, as well as establish high-valued utilization technologies for deep-sea biological resources?

3.2.4 Marine Security

The ocean is a complex nonlinear system that includes interactions among multi-scale dynamic processes such as large-scale circulation, meso (and sub-meso) scale eddies, small-scale internal waves, and micro-scale mixing. On one hand, the oceanic multi-scale dynamic processes have a direct impact on the underwater and surface navigations. On the other hand, through modulating the marine hydrological environment, the multi-scale dynamic processes can alter the acoustic, photoelectric, and magnetic fields, which further influence the underwater detection, communication, navigation, and target recognition in the ocean (Colosi and Worcester 2020). Therefore, it is an urgent scientific problem to reveal the oceanic multi-scale dynamic environment and its effects on various physical fields and through which to make their characteristics "transparent".

The extreme deep-sea environment has fostered the development of unique gene structures and physiological mechanisms in marine microorganisms. With the increasing of human activity in the oceans, some pathogenic microorganisms in the deep ocean would be released into the human environment, which could result in the emergence and spread of new and unknown microbial pathogens on land and in offshore areas, causing incalculable harm to human health, social stability, and ecological security. Explorations of deep-sea pathogenic microorganisms and strategies for deep-sea microbial infection prevention, control, and biosecurity are therefore important and frontier researches.

3.3 Deep-Space Resources and Deep-Space Economy

3.3.1 Exploration, Development, and Utilization of Space Resources

Space resources are defined as material or non-material resources from outer space, including location resources, environmental resources, and mineral resources, which can be developed and utilized by humans to deliver economic and other benefits. For the future sustainable development of society, scientists must come to understand the types, scale, distribution, and formation mechanisms of space resources. Non-material space resources must be effectively utilized, and material resources developed, so methods and technologies must be developed for their enrichment, mining, storage and transportation.

Research on space resources encompasses many basic fields. In order to effectively utilize space resources, and to develop our capability for space exploration, we will have to develop quantitative, high-resolution remote sensing and in-place detection technologies, expand the use of terrestrial, lunar, and planetary orbital space, and establish observation, communication, navigation, and positioning infrastructure across the solar system. We must establish and maintain base stations on the surface of the moon, the planets, and their satellites. To service these bases, we must develop in-situ 3D printing capabilities, establish water, hydrogen and oxygen caches, utilize local mineral resources, and introduce other related technologies. By utilizing in-situ resources, lunar or planetary bases will provide convenient and low-cost access to deep space. Technologies for advanced mineral mining, extraction, and purification, as well as safe storage and transportation, will be developed.

3.3.2 Utilization and Transmission Technology for Deep-Space Solar Energy

Use of the abundant solar energy in space, and other space resources, is critical for exploration of deep space, as well as for the promotion of sustainable development of society and the national economy here on Earth. In order to guarantee safe, green, sustainable energy as space activity increases, and to construct large-scale energy plants in space, the following key issues must be addressed: Solar power plants should be developed, based on thermal cycles, with working technologies such as radiation heat exchangers and cooling systems to make them suitable for operation on stratospheric platforms or in a lunar environment. Breakthroughs should be sought in high-efficiency solar cells and key materials preparation for space-based solar power stations. Structural design and control systems for high conversion efficiency solar cells, and preparation of photocatalytic materials for high efficiency hydrogen production using lasers should be explored and the general service performance of materials in a space environment evaluated. We must reach a better understanding of fluid and materials interactions and related system mechanics under the effects of high-intensity lasers, develop experimental methods for working with materials and structures under complex loads, develop multi-physical field-test technology, and explore the basic laws of the behavior of materials and structures under the action of high-intensity lasers. The basic data thus obtained will provide theoretical support for microwave radiation or laser energy transmission in near space.

3.3.3 Multi-scenario Economic Development in Space

'Space economy' refers to the technologies, products, services, and markets created by space-based and space-related activities, including economic benefits derived from the exploration, development, and utilization of space, such as space manufacturing, space agriculture, space resource utilization, space energy, space tourism, space culture industries, and space support and maintenance services. The space economy is a completely new form of business. Space resources are effectively limitless and have enormous potential for human exploitation. Space between the Earth and the moon is clearly the main field for exploration and is also the primary strategic space available for development of the aerospace industries of various countries in the future. The development of a space economy will be a new source and engine of economic growth, led by aerospace technology and driven by ingenuity and innovation. The development of diverse scenarios for a functioning space economy will promote government and social investment, drive innovation in traditional industries, promote the emergence and growth of new industries, stimulate economic growth, and be of enormous long-term benefit to human society.

3.3.4 Theories and Technologies for Human Intervention in the Trajectories of Dangerous Small Celestial Bodies

To make peaceful use of space and ensure the survival of human civilization, we must consider the harm that could be caused by the impact of small near-Earth objects on the planet's surface, not least the considerable quantities of space debris from mankind's own efforts in space over the past sixty years that are currently circling the Earth in orbits with varying degrees of stability. Theoretical research on human intervention in the trajectories of celestial objects requires focus on the following technical problems: ① development of detection methods for fast-moving 'dark'

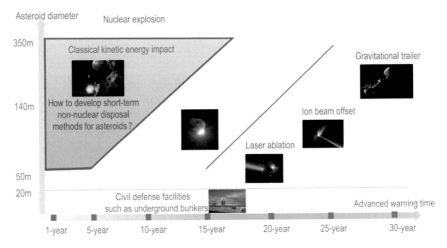

Fig. 3.3 Applicability of existing planetary defense methods

space targets, ② how to achieve large-scale, long-distance, non-contact, and efficient space debris removal, ③ how to scientifically and rationally plan the removal of space debris to obtain maximum benefit, and ④ how to utilize space waste and turn it into useful commodities (Fig. 3.3).

3.4   Mechanisms, Prediction, and Prevention of Natural Disasters

3.4.1   Earthquake Prevention and Mitigation of the Effects of Strong Continental Earthquakes

The formation and evolution of Earth habitability have been accompanied by many catastrophic events, including earthquakes. Most earthquakes occur in the upper and middle crust and can result in huge casualties and economic losses. A better understanding of the geophysical environments, stress states and dynamic processes in the focal depths of great earthquakes will lay a solid foundation for studies of earthquake preparation, rupture process and disaster mitigation.

The continental China has a highly complicated seismotectonic background. Great earthquakes occur frequently, mainly controlled by contractions among plate, intraplate mantle convection, and interaction between active tectonic blocks. Nevertheless, the complexity of seismotectonic structures and dynamics, the diversity of deformation and kinematics, and the randomness of temporal and spatial distribution of great continental earthquakes make it difficult to reach a full understanding of the factors resulting in earthquake occurrence. Two aspects of study are essential: mechanism of earthquake generation and effective earthquake early warning.

To study the mechanisms of great continental earthquakes, we need to bring modern science and technology to bear in monitoring the temporal and spatial evolution of tectonic stress in seismic zones. We need to fully comprehend the processes of energy accumulation and release in active fault zones and analyze the deep and shallow structures of fault zones and their relationships with deep ductile loading of faults and stress accumulation in seismogenic zones. Combining this with studies of fault friction and rheological experiments will reveal the generation processes of great earthquakes, the nucleation of earthquake ruptures, the characteristics of co-seismic slip and post-earthquake deformation. These studies will enhance a comprehensive understanding of the physical processes and temp-spatial evolution of earthquake occurrences.

At present, early warning systems are useful in mitigating the destructive disasters of earthquakes, but they could be improved enormously. For the future, there are two key directions: first, improvement of existing early warning theories and methods, including the development of new earthquake early-warning algorithms using the three-dimensional accelerometers now commonplace in smart mobile phones in conjunction with big data and artificial intelligence technology; second, to explore earthquake early warning systems based on the electromagnetic waves associated with earthquakes, or the elastic, light-speed gravity waves generated by seismic dislocation, to enable more rapid detection than the current systems based on detection of seismic waves.

3.4.2 Observation, Mechanism and Possible Control of Artificially Induced Earthquakes

'Induced earthquakes' refer to seismic activity induced or triggered by human activity. Increasingly frequent and serious induced earthquakes are being generated by human exploitation of natural resources, particularly the construction of high dams and large-capacity reservoirs, the development and utilization of deep underground space, and large-scale shale oil and gas exploitation. Studies of induced earthquakes should focus on three areas: (1) continuous observation using dense surface and underground seismic arrays and optical fiber strain sensors, and developing big data processing technology based on artificial intelligence to achieve continuous detection and monitoring of micro, medium, and strong earthquakes; inversions of stress fields, fluid pressure fields, three-dimensional fine velocity structure, and tomography using the continuous observation data to improve on-site real-time detection and monitoring for signs of fault activation; (2) experimental researches on faults with different lithologies, diverse scales, and various levels of maturity to understand the spatio-temporal processes of fault activation and earthquake induction, especially the nucleation, expansion, and arresting mechanisms of seismic and aseismic fractures of faults with varying maturity under fluid action; (3) to transform existing well sites and wastewater reinjection sites where induced earthquakes have occurred into experimental sites to carry out advanced experimental researches on the controllability of induced earthquakes to reduce the earthquake magnitude.

3.4.3 Marine Hazard Forecast and Prediction

The ocean breeds the most extensive devastating meteorological and geological hazards on Earth in direct or indirect ways. Although humans have made prominent progress in forecasting and predicting marine hazards, understanding of their generation mechanisms still needs to be improved. For meteorological and climatic hazards, such as typhoons and extreme EI Niño, the key factors to improve their forecast and prediction capabilities are to clarify the impact of small-scale and mesoscale ocean dynamic processes on thermal structure of the upper ocean, to develop accurate parameterization for air-sea turbulent exchange, and to establish a high-resolution Earth system model associated with a multi-element coupled assimilation system. For geological hazards, such as earthquakes, landslides, and tsunamis, to better understand their controlling geological and geophysical generating conditions, it is necessary to conduct various underwater surveys (e.g., geologic and geophysical observations, ocean drilling) to identify fault structures, rock compositions, stress characteristics, temperature and pressure conditions and so on; Then the controlling factors of geological disasters can be revealed by combining observations with laboratory experiments and numerical simulations. The urgent problem to be solved is to establish a scientific early warning system to mitigate disasters. Based on theoretical research, identifying high-risk areas and deploying submarine observation networks can monitor submarine disasters in real time and obtain key physical parameters, which will significantly improve the early warning capability of the geological disasters.

3.5 Ecological Safety

3.5.1 Ecosystem Structures and Processes

An ecosystem is the set of interactions between a community of biological organisms and its physical environment. Ecosystem structure is the basic attribute of ecosystems, reflecting how various elements in the ecosystem interact with each other, and also determining ecosystem functions and the ecosystems' responses and adaptation strategies to the external environment. Ecological processes include the ecosystem material cycle and energy flow. The ecosystem material cycle is the process by which chemical elements circulate between the biological population and the inorganic environment. Energy flow is the process by which the stored solar energy of green plants is transferred unidirectionally along the food chain, decreasing gradually as it goes. The material cycle and energy flow account for the interactions and relationships between biotic and abiotic elements within ecosystems and among the biotic factors of organisms at various food chain levels, representing the most important functions of an ecosystem. It is pivotal to understand the responses and adaptation mechanisms of ecosystem structures and processes to global change. Quantitatively revealing how ecosystems respond and adapt to global change will provide the scientific basis for human adaptive intervention in ecosystems. In order to maintain Earth's habitability in the future, three issues need to be addressed: ① Identification of the mechanisms governing ecosystem structure and stability, ② determination of the material cycle and energy flow mechanisms of ecosystems, and ③ quantitative assessment of ecosystem responses to global change.

3.5.2 Soil Health

With the increasing global population, the demand for food is rising inexorably day by day. Soil health plays an important strategic role in global food security. However, high-intensity human activities are causing severe degradation in soil quality of cultivated land, poor soil health, and continuous decline in grain production capacity. The *Status of World's Soil Resources Report* suggests that soil worldwide faces issues including acidification, nutrient imbalances, salinization, pollution, erosion, hardening and compaction. The ecological services and functions of soil are declining, which seriously restricts its capacity to sustain food supply. To improve the fertility and soil quality of cultivated land and maintain healthy soil, we must address the following issues: ① Improving the soil fertility of cultivated land to support global food production with a safe margin to avoid shortages; ② Preventing, controlling, and repairing soil pollution on cultivated land to ensure food safety and proposing strategies for the safe and sustainable use of contaminated soil to ensure soil health and food quality; and ③ making use of the ecological services and functions of soil organisms in cultivated land to maintain sustainable soil health and food security, studying the multitrophic biological interactions, immunity and detoxification functions of soil organisms to maintain soil ecosystems to ensure soil health and long-term food security.

3.5.3 Environmental Pollution Control

With the rapid industrialization and urbanization, we are now facing increasingly serious problems of environmental pollution. Although many nations have embarked on research programs to examine the ecological and social effects of environmental change and pollution, these initiatives have so far failed to solve the fundamental problem of ensuring sustainable human development. The following are urgent scientific questions for the future. First, we must enhance research on the transmission, migration, sedimentation, and distribution of environmental pollutants, and on regional environmental processes and risks at the global scale under climate changes. Meanwhile, we must analyze viruses' origins, transmission, and monitoring and their mechanisms based on environmental pollution processes. Second, we must strengthen research on the ecotoxic effects of pollutants and their transformation products and explore the impact of environmental pollution on ecosystems. We must conduct ecological risk assessments, develop early warning systems, and predict the ecological effects of future environmental changes. This will allow us to establish a complete theoretical system for environmental biology research and understand the changes and interaction mechanisms in the Earth's biosphere.

3.6 The Carbon Cycle and Carbon Neutrality

Carbon dioxide (CO_2) is the most important anthropogenic greenhouse gas. Carbon neutrality refers to the global or regional removal of anthropogenic CO_2 by energy conservation, emissions reduction, afforestation, and carbon capture, so as to reach net-zero carbon emissions. However, it should be noted that CO_2 is not the only anthropogenic greenhouse gas. The Paris Agreement of 2015 proposed a balance between greenhouse gas emissions caused by human activities and their removal in sufficient quantities to reach a net-zero emissions position in the second half of the twenty-first century. An IPCC Special Report on Global Warming in 2018 pointed out that the maximum 1.5 °C global temperature rise specified in the Paris Agreement will require global CO_2 emissions to reach a net-zero level around 2050, and that net-zero emissions of non-CO_2 greenhouse gases needs to be achieved before then (IPCC 2018).

The essence of carbon neutrality is the interaction of the Earth's climate system with the carbon cycle. The Earth's climate system is composed of the lithosphere, atmosphere, hydrosphere, cryosphere, and biosphere, which together regulate the Earth's temperature, precipitation, and air pressure. The principal factors controlling the change of the Earth's climate, such as land surface temperature, vary on different time scales. In a temporal scale of over a million years, the tectonic cycle is the main driving force, whereas on a scale of one hundred thousand years, the most important factor is the Milankovich cycle. Note that the response of the Earth's climate to changes in solar radiation intensity is complex and non-linear. Other factors influencing the Earth's climate include marine physical processes (such as sea surface temperature oscillations in tropical oceans) and the chemical composition of the atmosphere (such as the content of aerosols and greenhouse gases including CO_2), which play a significant role on relatively short time scales. The carbon cycle involves the migration and transformation of carbon and its compounds between and within the atmosphere, land, ocean, and other spheres. Its core encompasses cross-sphere and multi-scale carbon fluxes, processes, mechanisms, and interaction with the climate system. The injection of anthropogenic CO_2 into the atmosphere and its subsequent distribution across the ocean, land and atmosphere components of the Earth system have led to an unprecedented perturbation of the global carbon cycle that has been ongoing since the Industrial Revolution (Gruber et al. 2019). Since carbon dioxide (CO_2) is the most important anthropogenic greenhouse gas, carbon neutrality generally refers to net-zero CO_2 emissions globally or regionally attained by balancing the emission of CO_2 with its removal through energy conservation, emissions reduction, afforestation, and carbon capture. However, CO_2 is not the only anthropogenic greenhouse gas. The Paris Agreement defined climate warming targets of 2 °C, 1.5 °C, and 1v, associated with end-of-century atmospheric CO_2 concentrations of ca. 450, 400, and 350 ppm, respectively. These scenarios approach carbon neutrality by about 2070, 2055, and 2040, respectively, and remain negative thereafter (Hansen et al. 2017). More than 130 countries have proposed commitments on the reduction of individual carbon neutrality-related emission. Adoption and implementation of the actions proposed by many countries will introduce another set of perturbations to the global carbon cycle on decadal time scales.

Research should focus on increasing the precision of carbon flux estimates and revealing the controlling processes and mechanisms to enable more accurate predictions of future climate changes. In order to achieve carbon neutrality in response to climate change, it is vital to examine how the global carbon cycle evolves under both natural conditions and human intervention, as well as the response and feedback of the Earth's climate system.

Emissions reduction (reducing CO_2 emission into the atmosphere) and increasing carbon sinks (enhancing the removal of atmospheric CO_2) are two essential pathways for realizing carbon neutrality. When compared to the extensive researches on the paths, measures, and policies related to CO_2 emissions reduction, the patterns, time scales, and future trends of carbon sinks in various spheres and their interactions with the climate system remain to be better constrained. In addition, increasing attention is paid to a variety of methods targeting the alleviation of climate change, or climate intervention, but its impact on the carbon cycle and ecosystem is not clear and therefore requires more research.

In the context of carbon neutrality, the following research is seen to be essential.

3.6.1 Cross-Sphere and Multi-scale Processes and Mechanisms of the Carbon Cycle and Their Relationships with the Climate System

The core of the carbon cycle is multi-scale carbon fluxes, processes and mechanisms, which involves the migration and transformation of different forms of carbon within and between spheres including the atmosphere, land and ocean. Changes in the temporal and spatial patterns of carbon fluxes determine the degree and scope of contemporary climate change. There have been increasing observations of global and regional carbon fluxes and the general pattern of the CO_2 source and sink in each sphere. Whereas there is still a lack of supporting data and mechanistic understanding of the evolution of carbon spatiotemporal patterns under the combined stress of climate change and human activity, it is therefore necessary to accurately simulate and predict the cross-sphere and multi-scale processes of carbon evolution, which provides a solid theoretical basis for the path towards carbon neutrality.

3.6.2 Budgets, Reservoir Capacity, Uncertainty, and Evolution Trends of Carbon in Terrestrial, Oceanic, and Land-Sea Coupled Systems

About 46% of anthropogenic CO_2 remain in the atmosphere. The other 54% rapidly enters the oceanic (23%) and terrestrial (31%) ecosystems by sea-air and land-air exchanges, respectively. The core issue in carbon cycle research is therefore to accurately understand the budgets, reservoir capacities, and dynamics of carbon in terrestrial and marine ecosystems. Terrestrial ecosystems work as an efficient carbon sink, absorbing $3.4 \, \text{Gt C yr}^{-1}$ from the atmosphere (2010—2019), of which carbon storage in vegetation and soil is 2241 Pg C. With increasing anthropogenic CO_2 emissions, the CO_2 absorbed by terrestrial ecosystems has increased at a rate of 0.39 Gt per decade (1960–2019). Carbon fixation operates in a number of ways in terrestrial ecosystems, including photosynthesis by vegetation and organic matter secreted by plants roots into the soil, and the preservation of dead and decomposed plants in the soil. Due to the influence exerted by climate change and human activity, there is still great uncertainty regarding the capacity of terrestrial ecosystem in sequestering anthropogenic CO2 on both regional and global scales. The stability and sustainability of the carbon reservoirs also need to be better constrained. Future research foci are: ① the capacity of carbon in different forms and the uncertainty, stability, and sustainability of the carbon sink in typical terrestrial and marine ecosystems ② the spatiotemporal distribution patterns and evolution of carbon reservoir under the combined influence of natural processes and human activities; ③ the roles terrestrial and marine carbon sinks play in different carbon neutralization pathways.

3.6.3 Impact of Carbon Neutrality on the Coupling System Between Carbon Cycling and Climate

Achieving carbon neutrality will involve multi-spheric interactions between the ocean, land and atmosphere and human beings. How humans affect the Earth is crucial to the sustainable development of nature, society, and the economy. Studying the impact of carbon neutrality on the coupling between carbon cycles and climate is a new field in geoscience.

3.6.4 Scientific and Technological Basis for Negative Emissions Technology

Negative emissions technology (i.e., increasing carbon sinks) is vital for achieving carbon neutrality. At present, increasing carbon sinks on land by for example afforestation is an internationally recognized approach. Oceanic carbon sinks (such as those enhanced by marine fertilization) have great potential towards creating negative emissions, but relevant research is scarce. On the basis of accurately monitoring natural and anthropogenic carbon sources and sinks, it is necessary to reinforce basic research on negative emissions technologies including geoengineering, so as to build a conceptual framework for an optimal path towards carbon neutrality.

In order to monitor carbon sources and sinks, a new monitoring, verification, and support (MVS) system is needed. The horizontal transport flux of carbon within a sphere (e. g., land-river) and across interfaces (e. g., river-estuary-coastal ocean-open ocean) needs to be better quantified to reduce its impact on vertical carbon flux assessments. Based on high-quality field data of carbon sources and sinks, the Earth system model can be optimized to predict possible future scenarios and uncertainties in the context of carbon neutrality.

Basic research is critical for implementing any geoengineering project targeting the enhancement of carbon sinks based on natural ecosystems. To achieve carbon neutrality, we first need to construct high resolution data products of regional carbon budgets and develop models and theories to accurately evaluate the sink potential of land, coastal zones, and marine ecosystems. A comprehensive feasibility assessment system should be established for negative emissions technologies, and innovative theories and technologies should be explored. Finally, we need to integrate basic research with applied technology to develop carbon neutral solutions and plan climate change countermeasures based on natural ecosystems. In this way we will strive to protect the biodiversity of our environment and generate synergistic benefits for mankind and for nature as a whole.

References

BP (2020) BP statistical review of world energy. https://www.bp.com/en/global/corporate/energy economics/statistical-review-of-world-energy.html

Colosi JA, Worcester PF (2020) A seminal paper linking ocean acoustics and physical oceanography. J Acous Soc Am 148(6):R9–R10

Daletos G, Ebrahim W, Ancheeva E et al (2018) Natural products from deep- sea-derived fungi–a new source of novel bioactive compounds? Curr Med Chem 25(2):186–207

Future Earth Transition Team (2012) Future earth: research for global sustainability—a framework document, 199

Gruber N, Landschützer P, Lovenduski NS (2019) The variable southern ocean carbon sink. Ann Rev Mar Sci 11:159–186

Hansen J, Sato M, Kharecha P et al (2017) Young people's burden: requirement of negative CO_2 emissions. Earth Syst Dyn 8(3):577–616

Hein JR, Mizell K, Koschinsky A et al (2013) Deep-ocean mineral deposits as a source of critical metals for high- and green-technology applications: comparison with land-based resources. Ore Geol Rev 51:1–14

Hein JR, Koschinsky A, Kuhn T (2020) Deep-ocean polymetallic nodules as a resource for critical materials. Nat Rev Earth Environ 1:158–169

Hochella MF (2008) Nanogeoscience: from origins to cutting-edge applications. Elements 4(6):373–379

IPCC (2018) Special report on global warming of 1.5 °C. Cambridge University Press, UK

Jin ZJ, Hu WX, Zhang LP et al (2007) Deep fluid activity and hydrocarbon accumulation effect (in Chinese). Science Press, Beijing

Li SZ, Suo YH, Dai LM et al (2010) Development of the Bohai Bay Basin and destruction of the North China Craton (in Chinese). Earth Sci Front 17(4):64–89

Li WX, Hong J, Chen B et al (2019) Distribution regularity and main scientific issues of strategic mineral resources in Central Asia and adjacent regions (in Chinese). Bull Natl Nat Sci Found China 33(02):119–124

Liu JW, Zheng YF, Lin HY et al (2019a) Proliferation of hydrocarbon-degrading microbes at the bottom of the Mariana trench. Microbiome 7:47

Liu QY, Zhu DY, Meng QQ et al (2019b) The scientific connotation of oil and gas formations under deep fluids and organic-inorganic interaction (in Chinese). Sci China Earth Sci 62:507–528

Meng QR (2017) Development of sedimentary basins in Eastern China during the yanshanian period (in Chinese). Acta Metall Sin 36(4):567–569

Stocker TF, Qin D, Plattner GK et al (eds) (2014) Climate change 2013: the physical science basis. In: Contribution of working group I to the fifth assessment report of IPCC the intergovernmental panel on climate change. Cambridge University Press, Cambridge

Sun CY, Tan X, Zhou XY et al (2003) A research review of mineral materials made of oceanic polymetallic nodules and cobalt-rich crusts (I) (in Chinese). Metallic Ore Dress Abroad 9:4–11

Sun WD, Ling MX, Wang FY et al (2008) Pacific plate subduction and mesozoic geological event in eastern China (in Chinese). Petrol Geochem 27(3):218–225

Tortorella E, Tedesco P, Esposito FP et al (2018) Antibiotics from deep-sea microorganisms: current discoveries and perspectives. Mar Drugs 16(10):355

UNESCO (2019) The United Nations world water development report 2019: leaving no one behind. UN, New York

Xiao WJ, Song DF, Windley BF et al (2019) Research progresses of the accretionary processes and metallogenesis of the Central Asian orogenic belt. Sci China Earth Sci. https://doi.org/10.1007/s11430-019-9524-6

Xue CJ, Zhao XB, Zhao WC et al (2020) Deformed zone hosted gold deposits in the China-Kazakhstan-Kyrgyzstan-Uzbekistan Tian Shan: metallogenic environment, controlling parameters, and prospecting criteria (in Chinese). Earth Sci Front 27(2):294–319

Zhang LF, Tao RB, Zhu JJ (2017) Some problems of deep carbon cycle in subduction zone (in Chinese). Bull Mineral Petrol Geochem 36(2):185–196

Zhu RX, Xu YG (2019) The subduction of the west Pacific plate and the destruction of the North China Craton (in Chinese). Sci China Earth Sci 62:1340–1350

Zhu WL, Wu JF, Zhang GC et al (2015) Discrepancy tectonic evolution and petroleum exploration in China offshore Cenozoic basins (in Chinese). Earth Sci Front 22(1):88–101

Chapter 4
Scientific and Technological Support: Fundamental Theoretical Issues with Revolutionary Technologies

4.1 Geophysical Exploration Technologies for Deep Earth

4.1.1 Theories and Technologies for Seismic Surveys

Development of sea floor detection and distributed optical fiber sensing seismic survey technologies

Global-scale collaborative observation on land and sea requires cutting-edge deep-Earth research. The fundamental prerequisite for obtaining full information in the deep Earth is to conduct high-quality seismic observation in marine areas, given that the oceans and seas cover 70% of the Earth's surface. In recent years, intensive research and development (R&D) have made effective submarine observation possible by supporting a series of deep geophysical exploration projects which implement broadband ocean bottom seismography (OBS) (PLUME, SAMPLE, Pacific Array and others). Meter-scale observation density of seismic waves has been achieved using distributed optical fiber sensing technologies, which find wide application in petroleum exploration as well as in natural earthquake observation and are gradually established as the new generation of seismic survey technologies. These technologies make long-term and continuous seismic surveys possible in the deep seas, sub-zero zones, and even on extraterrestrial bodies, all of which are difficult to cover using traditional networks of seismographic stations. The introduction of new technologies (e.g. big data, cloud computing, and artificial intelligence) may support the development of new imaging methods based on distributed optical fiber sensing. Data with TB and even PB volumes from large-scale, high-density surveys, are available for processing using big data mining, which is driving a revolution in seismic survey. Meanwhile, the construction of data sharing platforms ensures maximum utilization of seismic survey data.

© Science Press 2022, corrected publication 2022
The Research Group on Development Strategy of Earth Science in China,
Past, Present and Future of a Habitable Earth, SpringerBriefs in Earth System Sciences,
https://doi.org/10.1007/978-981-19-2783-6_4

Emphasis on breaking through survey system limitations, and drawing lessons from multi-disciplinary thinking

Limited by its reliance on surface systems, deep exploration has to rely on simple telemetry to send and receive data. Uneven distribution of seismic stations, as well as the occurrence of earthquakes, constrains the accuracy and resolution of deep Earth structural imaging. Breaking through these technical bottlenecks is an important objective for future deep exploration studies. For example, data from current observation systems have to be supplemented and expanded by extracting structural signals from continuous noise records, previous seismic signals, and data using wave field interference. In wave field interference (or continuation) technology, surface observation systems can be migrated to regions near deep study targets, which makes 'telemetry' become 'short-distance measurement', significantly improving structural imaging accuracy and resolution. However, the fields of seismology, interference theories, signal extraction, and analytical technologies are still waiting for the effective application of these concepts and methods in actual deep exploration.

Enhanced survey methods for the lower mantle and core, and promotion of comprehensive, quantitative, and multi-scale research

Study of the structure of the lower mantle and core is currently weak and must be strengthened to achieve the objectives of understanding deep-Earth processes and the habitable Earth. Deep exploration will be required to accomplish a relatively unified and complete understanding of the Earth's systems from the deepest zones to the surface. Study of deep structures and properties need more precision to reduce uncertainty. Comprehensive understandings of deep structures, physical and chemical properties, dynamic processes, deep-shallow coupling, and sphere interaction, require more precise geophysical exploration and multi-disciplinary observation (e.g. geophysics, geology, rock-geochemistry). Dynamic simulations and experiments should also be carried out as part of a comprehensive, systematic, and multi-scale research program.

4.1.2 Quantum Sensing and Deep Geophysics

Quantum sensing gravity and magnetic technologies can produce results several orders of magnitude more accurate and fine-grained than those of traditional measurement methods (Peters et al. 2001). If the measurement precision of quantum gravity can be increased to 10^{-1} uGal, it may be able to record surface gravity changes caused by motion of the core (at present it is thought that surface gravity changes of approximately 0.1 uGal can be generated by decade-scale movements). Technological innovation in quantum sensing gravity and magnetic measurement means that we will soon be able to achieve high-accuracy observation of delicate structures in the deep Earth, their characteristics, and their dynamic variations. The mechanisms

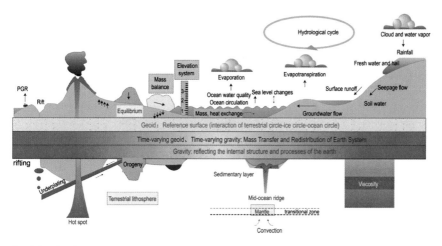

Fig. 4.1 Material distribution and transfer of geodetic datum and Earth system

of medium- and long-term variations in the inner core will be revealed and understood by acquiring new evidence from deep core-mantle coupling, movement patterns of the solid and liquid cores, and density differences at the core-mantle boundary. The patterns of Earth sphere interactions on different spatiotemporal scales will be determined by the acquisition of new information about inner Earth mass transfers and crust movements. The new technology will also immeasurably improve our capacity for monitoring and predicting earthquakes and volcanic activity, quantitatively monitoring mass transfers and changes in surface systems (e.g. continental water, sea water and glaciers), and evaluating the influence of climate changes and human activity on water resources (Fig. 4.1). Improvement of the accuracy and resolution of quantum geomagnetic measurement will also be helpful in achieving a better understanding of the motion of the liquid core, and for further exploring the mechanisms of decade-scale Earth variations.

4.2 Deep-Earth Geochemical Tracker and High-Precision Dating Techniques

To better understand the formation and evolution of habitable Earth, new technics are needed to constrain Earth's differentiation processes, develop new proxies that are sensitive to Earth's habitability and their variation rates, timelines of major geological events. Geochemical tracing methods and high-precision dating technics will play vital roles.

4.2.1 Geochemical Technologies for Tracing Early-Stage Earth Evolution and Core-Mantle Differentiation

'Early-stage' Earth evolution refers to important geological events that occurred in the range of ca.1 Ga—from planet formation to crust formation (4.56–3.6 Ga)—including primitive Earth materials composition and sources, giant impacts and their effects on Earth's composition, core-mantle differentiation, core formation and evolution of the geodynamo, crust-mantle differentiation, the ancient magma ocean, primitive continental crust, early-stage Earth structure types, timing of the emergence of pre-plate tectonics and plate tectonics, formation of the early-stage Earth hydrosphere and atmosphere, life origins, and other cutting-edge issues. However, material samples from early-stage Earth are extremely rare and valuable, and current traditional geochemical analysis technologies are inadequate to understand these ancient processes and features. Innovative new theories and technologies for isotope dating and geochemical tracing are urgently required and should be based on multidisciplinary cooperation. Time is the basis for studying geological processes. Besides traditional radioactive isotope dating methods (e.g. zircon U–Pb and individual mineral $^{40}Ar/^{39}Ar$), short half-life extinct nuclide dating has considerable advantages in mapping the chronology of early-stage Earth geological events. High-precision extinct nuclide dating (e.g., $^{53}Mn-^{53}Cr$, $^{26}Al-^{26}Mg$, $^{182}Hf-^{182}W$ and $^{146}Sm-^{142}Nd$ systems) accurately measures the differential timing of metallic core and silicate mantle, and of crust and mantle, and can also trace core-mantle differentiation and heterogeneous mantle processes. Development of analytical theories and technologies for high-precision dating of stable metal isotopes (Ti, V, Zr, Ni, Al, Ca, Cr, V, Zn, Cd, In, Pd, Os, Ru, etc.) can reveal the element distribution characteristics and controlling factors of high-temperature isotope fractionation in the process of core-mantle differentiation (e.g. temperature, pressure, oxygen fugacity, and chemical components), which is helpful for throwing light on the process of core-mantle differentiation and the formation mechanisms of mantle heterogeneity in early-stage Earth.

During core-mantle differentiation, relatively few light elements found their way into the core, and their separation under HPHT conditions following core formation could have provided the force required for geomagnetic field initiation. The development of high-precision analytical technology for light element isotopes (Si, C, S, N and O), particularly isotopes with low abundance ($\triangle^{17}O$ and $\delta^{34}S$), is essential for understanding the distribution behavior of light elements during core-mantle differentiation as well as the process itself.

4.2.2 Index System of Earth Habitability Elements

Understanding the evolution of factors making Earth habitable over time is a prerequisite for study of the formation of Earth habitability and the controlling mechanisms of the deep Earth; and is a considerable basis for comprehending the essence of every crucial geological event, their systemic driving forces, the connections between various system circulations, and their feedback mechanisms. Only once we have this information can the story of habitable Earth be fully told and the reasons for Earth's habitability dissected, which will finally establish the basis for constructing a systematic theory of habitable Earth. Nevertheless, there are many unresolved questions about the formation and mechanism of Earth habitability till nowadays. As we know, temperature, water, and oxygen are three essential elements of habitable Earth, their relative stability or dynamic balance in surface and near-surface systems are the key to maintaining Earth habitability. Early-stage life consisted mainly of bacteria, archaebacteria and other microorganisms, but it is still unclear how these microorganisms interacted with Earth's environment and how they contributed to creating the current configuration of these three basic elements. Concerning the Great Oxidation Event, neither the initiation nor the termination have been clearly explained, and even the timing of its episodes has not been accurately dated. Its temporal relationship and position in evolutionary sequence relative to a series of other vital events (continental crust evolution (growth, composition abrupt changes and uplifting), the global glaciation, flourishing of cyanobacteria, buildup of banded iron formations) have still not been established. Essentially, none of its interactions and correlations with other major events are fully understood. There are many methods for water body temperature reconstruction (e.g., oxygen isotopes in calcareous organism (foraminifers, bivalves, brachiopods, etc.), foraminifer Mg/Ca, coral skeleton Sr/Ca, foraminifer δ^{44}Ca and TEX86), but all of these indices have significant shortcomings. For instance, both the oxygen isotope method and the trace element method show deviations because the pH and salinity of water bodies affect the composition of isotopes and trace elements in planktonic foraminifer shells. This shortcoming will hinder the confirmation or rebuttal of scientific hypotheses, and also limit our ability to reconstruct the evolution of the habitable oxygen-bearing atmosphere and predict its future. An index system therefore should be constructed starting from the critical elements of habitable Earth and the necessary material basis for life, this will be helpful for reaching a deeper understanding of the formation and evolution of our habitable environment and will also be a focus of future research.

4.2.3 High-Precision Dating Technique

Application of secondary ion mass spectrometry (SIMS) and laser albration-plasma mass spectrometry in geochronology has been a great success in China and reached a very ? level in the international standard. It enables precise in-situ analysis of isotope ratios in the micrometer to sub-micrometer scale, and thus markedly improved analytical efficiency. Combined application of color spectrum-accelerator mass spectrometry technologies increases the accuracy of ^{14}C investigation capacity from milligram scale to molecular scale. To deploy the international Earth time scheme (EARTH-TIME), with the key objective of enhancing time resolution, the dating accuracy of U–Pb and Ar/Ar has been enhanced from roughly 1–0.1% and the differences between laboratories and between the previous dating systems have largely been obviated. This has facilitated the leap from merely determining the age of geological events to being able to trace the rates of progress of geological processes. By comparison, no significant breakthrough in high-precision ID-TIMS dating technology has been achieved in China. Core technologies for high-precision chronology are the development and calibration of spikes and standard samples, measurement of fundamental parameters such as decay constants, and the development of data processing software, all of which are now urgent requirements.

4.2.4 HPHT Experiments and Computing Simulation Techniques

As a bridge between geochemistry and geophysics, HPHT simulation represents a crucial backbone and foundation for elucidating and testifying new geoscientific theories, opinions and concepts. The future development of HPHT experiment and simulation techniques lies in combining geophysics, microscopic structural analysis, numerical calculation, and simulation. The simulation temperatures and pressures realized so far cover the conditions of crust to core. In particular, the combination of synchro with HPHT experiment technics paves new way to achieve significant breakthroughs in the study of deep Earth interior.

Geo-scientific research in China has always suffered from the shortcoming of emphasizing observation at the expense of simulation and integration. This situation is being improved recently asthe new generation of scientists is doing well, carrying out simulation experiments on dynamic processes (e.g., mineral phase changes, physical property testing, partial melting, and rheological properties) over the entire range of temperatures and pressuresof the upper mantle. Of particular notice is the diamond anvil cell (DAC) technology, developed at the Center for High Pressure Science and Technology Advanced Research (HPSTAR), is now capable of simulating conditions in the lower mantle and even in the core, which has allowed China to be a significant force in global research on the ultra-deep Earth. China's capabilities in other technologies also need to be strengthened, particularly technology of fluid studies, in-situ micro measurement under HPHT conditions, simulation of higher temperature and pressure conditions, and interdisciplinary integration in the use of HPHT simulation technology.

4.3 Deep-Sea Observation and Survey

4.3.1 Detection Technologies for Ocean Laser Profile

Understanding the subsurface vertical structures in the global ocean plays an important role in climate change research, underwater vehicles detection, and marine fishery development. However, mature technologies for collecting global subsurface water profiles with high-resolution and high-precision are limited at the moment. To achieve the full-parameter, high-resolution, integrated remote sensing detection of ocean profiles in a combination of marine optics, thermodynamics, dynamics, and ecology, it's necessary to study the mechanisms of high-sensitivity individual photons detection and identification, coupling between marine dynamics and optics, also coupling between sub-mesoscale surface topology and subsurface vertical structures.

4.3.2 Detection Technologies for Marine Neutrino

The scarcity of real-time observations of seafloor sediment and polar ice caps is a major bottleneck of the development of polar science. Neutrinos can penetrate seawater with relative freedom. They can also interact weakly with quarks of hydrogen and oxygen atoms in seawater, releasing charged particles and weak rays with extremely high velocities. Studying the mechanisms of the weak interactions between neutrinos and sea materials, the mechanisms of seafloor neutrino flux detection, and the telescope technology of oceanic neutrino, will enable the extraction of parameters from the deep sea, such as neutrino oscillation, mixing angle, and so on. Furthermore, this can help inverse deep-sea environment, detect the vertical sea ice structures, and explore the seafloor resources.

4.3.3 Deep-Sea and Transoceanic Communication Technologies

All wireless communication technologies relying on radio frequencies, optics and underwater sound are subject to certain limitations, making it extremely urgent to explore the marine information transmission technologies in the deep and open sea. The complex ocean environment requires constructing a new and fundamental theoretical system for satellite-based, high spatiotemporal resolution, thus assisting the investigation of the deep seas. This system involves the investigation of satellite-sea collaborative coupling communication mechanisms, laser-induced sound and acousto-optic transformation mechanisms, and underwater platform quantum communication patterns. The R&D aims to integrate laser, underwater sound, and quantum to build communication systems, and further breakthroughs are expected in these revolutionary new-generation technologies. These systems will establish the basis for developing deep-sea laser investigation satellites, neutrino detection satellites, and ocean quantum communication satellites.

4.3.4 Underwater Observation and Survey

The ultra-high-speed unmanned platforms that can satisfy stringent demands of underwater observation and survey are rarely reported both at home and abroad, while only the Russian "Poseidon" nuclear powered unmanned underwater vehicle has achieved a top underwater speed of around 200 km/h. Therefore, it is necessary to develop ultra-high-speed, unmanned, multi-state, invisible, and intelligent vehicles equipped with high-precision and intelligent navigation, positioning, and communication systems, as well as self-adaptive sensors for sampling and observation/survey. The onboard systems should have the functions of the survey, communication, and navigation by integrating underwater acoustic, optical, and quantum communication technologies. These vehicles will have the capacity for trans-media indefinite cruising at all weathers and depths, with self-sustaining and adaptive transformation technologies, which can realize rapid environmental awareness, machine learning, and artificial intelligence based on decision making.

4.3.5 Sea Floor In-Situ Surveys

Ocean drilling ships currently in service need to be equipped with high-capacity mud pumps and marine risers. Their enormous volumes and high operational costs may be crucial factors in explaining why the Moho has not yet been penetrated. A possible alternative technology for the future is placing drilling rigs on the sea floor, with power supplied from support ships on the surface. Drill cores would be collected and transported automatically, and the rigs equipped with ROVs for maintenance. The following technological systems must first be developed: (1) automatic AI drilling and exploration systems, (2) automatic coring technology, (3) core storage and transportation technology; and (4) sea floor anti-corrosion technology.

The crustal lithosphere, oceanic benthos layer, demersal hydrosphere, and atmosphere comprise a single interconnected unit. From the perspective of Earth systems it is vital to engage in long-term, in-situ observation of multispheric interactions (atmosphere, hydrosphere, biosphere and lithosphere) around the sea-atmosphere, sea-land, and sea floor boundaries (Fig. 4.2). Essential properties of in-situ, full-depth, topographic sensors are underwater calibration-free and high resistance to environmental damage. Long-term, stable, and comprehensive in-situ surveys of key physical and chemical parameters are required to combine sound, optics, spectrum, mass spectrum, and other technologies and methods. The areas that must be surveyed include material and energy transportation pathways (such as caverns created by

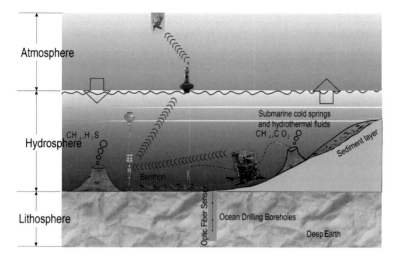

Fig. 4.2 Sketch of submarine long-term and in-situ observation on spanning-sphere

ocean drilling), deep-sea cold springs and hydrothermal fluids, and large-expanse oceanic physical–chemical fields. Breakthrough technologies for such a system include: comprehensive multi-parameter survey technologies for the sea-atmosphere, sea-land, and sea floor boundaries; quantitative analytical methods for extreme deep-sea environment; automatic intelligent power generation and management technologies for underwater environment; long-term, calibration-free, and stable operation technology for in-situ sensors; and automatic storage, protection, and long-distance and real-time transmission technology for batch data.

4.4 Deep-Sea Mobility and Residence

Deep-sea mobility is predicated on the development of ultra-high-speed, ultra-long distance, ultra-deep water, ultra-invisible and ultra-large size mobile platforms. The following key theories are required.

Ultra-high-speed: The relationship between microscopic (e.g. cavitation/Ultra-cavity growth, collapse and interaction) and macroscopic mechanisms must be established. The generation mechanisms of cavitation and cavity instability in navigational processes must be studied as well as the characteristics and mechanisms of cavity collapse. Physical models of collapse pressures should be constructed. The influencing mechanisms and patterns of ventilation on stability should also be studied along with the characteristics and mechanisms of ventilated cavity dropout (Wu 2016).

Super-cavitation drag reduction: Important considerations are the generation and annihilation mechanisms of vapor bubbles in water. The rules governing generation, variation and annihilation of vapor bubbles must be studied under low temperature and high-pressure conditions. The factors affecting drag reduction by bubble curtain caused by vapor should be identified. Fluid–gas–solid coupling mechanical models and simulation methods in high-velocity submarine navigation bodies must be studied. Revealing the drag reduction patterns of fluid-gas–solid three-phase coupling and the construction of simulations of resistance and controlling forces will provide strategies and algorithms for control in ultra-high velocity navigation.

Ultra-deep water: New materials must be developed with high durability, low density, abrasion resistance, and strong mechanical properties. These might include carbon fiber, ceramic materials, titanium alloys, other new nonmetals, and compound metallic materials with suitable mechanical and corrosion-resistance properties which will solve the problems of ultra-deep water pressurization and capsulation.

Ultra-invisible: Anti-noise mechanisms and denoising in cavitation states are essential. The empirical formulae for noise pressure and vibration velocity are deduced from the noise generated in the initiation and collapse stages of cavities. Noise in the entire process and its affecting factors must be examined and understood, from the initiation of the generation of free gas to cavity collapse. The study of denoising patterns starts from their affecting factors.

Ultra-large size: The dynamic properties and fluid–solid coupling mechanisms of ultra-large mobile platforms must be established, as well as the dynamic responses of ultra-large platforms navigating in complex ocean environment. The mechanical rules governing fluid–solid coupling in ultra-large mobile structures must be revealed. The interactions of nonlinear mechanical processes in ultra-large structures must also be studied.

The deep sea offers strategic opportunities for the expansion of the space available for human habitation (Fig. 4.3). Nevertheless, several crucial scientific problems must first be solved. Theories are required to predict the dynamic responses and intensity of complex ocean engineering under extreme loading conditions. Specific issues include fluid–solid coupling problems caused by thumping, green water, shaking, and other phenomena that affect machinery and structures in the extreme environment of the deep ocean will influence the engineering of human submarine residences; occurrence, evolution, and description of extreme ocean environment events; theories and methods for ocean structure/equipment response inhibition. Deep-sea environmental protection must also be taken into consideration, particularly the influence of human activity in underwater cities on marine ecology.

4.5   Exoplanets Exploration

The three basic conditions essential for the existence and survival of all life of Earth are energy, certain fundamental elements (C, H, N, O, P and S) and liquid water.

Fig. 4.3 Concept for a future underwater city. The entire city is constructed within an enormous transparent underwater bubble that will contain all the buildings and other structures normally found in terrestrial cities

The first two are commonplace throughout the universe. The third condition therefore becomes the key constraint for Earth habitability and also the most important reference index for the identification of habitable exoplanets (Meadows and Barnes 2018; Pierrehumbert 2010; Zhang 2020). Long-term existence and maintenance of liquid water on a planet's surface is dependent on three factors: the properties of the planet itself, the properties of its star, and the properties of its planetary system. The properties of the planet itself include planet size, volume and mass, its composition at the time of initial formation, internal structure, geological activity (volcanic activity, plate movement, heat plumes, etc.), magnetic field, planetary orbit (semi-major axis, eccentricity and inclination), rotation rate, atmospheric composition and quality, clouds and aerosols, etc. The crucial properties of its star are temperature, activation, metal abundance, rotation rate, and early-stage luminosity. Planetary system properties include interactions between planets, stellar gravitational and tidal action on planets, and the influence of other celestial bodies on the planetary system. Rapid development of exoplanet exploration will be beneficial from the enormous advances in investigation technologies, for example, wide-field photometry on planetary surfaces and high precision optical spectrum surveys. In the future, it will be essential to develop new theories and to carry out simulations, experiments, and observations to understand whether planetary surfaces can sustain liquid water over long periods of time.

4.5.1 Exploration of Atmospheres and Life on Exoplanets

Atmospheric observation theories for exoplanets are relatively mature. Whereas, technological measures are limited by the size of current telescopes. Coronagraphy and stellar shading board technology are applied for direct imaging. The former method is technologically mature but is untested in the observation of exoplanets and the latter is still in the R&D stage, being developed exclusively by NASA, and is currently being prepared for atmospheric observation of exoplanets (National Academies of Sciences, Engineering, and Medicine 2018).

Current observation technologies can only explore the atmospheres of planets that have large volumes and relatively abundant atmospheres, for example, 'hot Jupiters' and 'hot Neptunes'. It is very difficult to observe the atmospheres of terrestrial planets. The apertures of current telescopes are too small, limiting observation capabilities, and atmosphere detection on terrestrial planets is more complex. For instance, the effects of clouds on stellar light, and of our own atmosphere on atmosphere observation signals, present enormous challenges to observation precision. Coronagraphs, or stellar shading plates, deal with the problem of stellar light shading. However, with respect to cloud shade, ultra-long exposure times are required to improve signal-to-noise ratios sufficiently to solve the problem. Large-aperture space telescopes (several meters aperture) must therefore be developed specifically for the observation of exoplanets. In the stage of start-up and experience accumulation, observation of huge planets like hot Jupiters has been attempted with small-aperture telescopes (roughly 1 m).

The investigation of life-related gaseous components in the atmospheres of exoplanets is currently accepted as the most promising approach for searching for exolife. New technologies will be developed to achieve the investigation objectives and study will be carried out in combination with a full deep space investigation plan, which is expected to accomplish great breakthroughs in the future.

4.5.2 Climate Environment on Oceanic Planets

Liquid water on the Earth's surface accounts for roughly 0.02% of the total volume of the planet. On one type of exoplanet, water may account for 5% or even more of the overall volume, which means that the sea depths on this type of planet could be up to tens or even thousands of kilometers. Planets of this type are called ocean planets. If a planet is totally covered by ocean, there will be no weathering processes like those on Earth. Carbon dioxide from volcanic eruptions will therefore simply accumulate in the atmosphere. Ultimately, a runaway greenhouse effect is likely to develop, driving the atmosphere of an ocean planet into a fugitive state. In addition, questions relating to the distribution and intensity of ocean circumfluence, surface climate, and life supporting capability of ocean planets will require in-depth research.

4.5.3 Ocean Currents and Heat Transportation on Lava Planets

Observations indicate that the surface of a certain type of solid and extremely hot planet is a molten sea of magma, very similar to that of early Earth. The viscosity of magma is higher than that of aqueous ocean but it can still flow, redistributing heat energy and mass around the entire planet. The Spitzer and Hubble telescopes have both observed the characteristics of heat distribution on molten planets, especially those with tidal locking, where rock components may be molten on one hemisphere and consolidated on the other, exerting a huge influence on the configuration of the entire planet. Nevertheless, research in this area is scant at present.

Every country should coordinate its scientific and technological forces to seize international leading-edge themes for exoplanet detection, to develop large-size and high-precision space telescopes, to develop new technologies and methods, and to promote exoplanet detection. In China we have project CHES, which is searching for habitable terrestrial planets in habitable zones of the Sun's neighboring stars using astronomical observation methods. The super Kepler project is searching for Earth 2.0 using large-view, small-aperture arrays. The 1.2 m ExIST spectroscopic telescope is searching for relatively close exoplanets and is exclusively applied in the "Tianlin" (Chinese means Sky Neighbors) project to study planetary properties in habitable zones and search for exolife on suitable candidate planets. In addition, the spatial interference method is being used to search terrestrial planets in habitable zones and image them directly using large-scale space telescopes in the "Miyin" (Chinese means Voice Searching) project.

4.6 Infrastructure Framework and Theories of Positioning, Navigation, and Timing (PNT) Services

The traditional geocentric space–time datum maintenance technology and its corresponding PNT technology based on Global Navigation Satellite System (GNSS), suffers from inherent vulnerabilities, regionality and other practical limitations. GNSS signals can't explore deep space, penetrate water bodies, or provide high-precision and high-sensitivity observations for detecting the variations of deep Earth materials. Therefore, they could hardly meet the requirements of deep space, deep sea, and deep Earth exploration. The space–time datum based on geometric and optical observations can only provide PNT services in local areas. As a result, the new challenge for future space–time datum construction and maintenance is to seek technologies based on new physical principles supporting the research of "three deep and one system" and livable environment.

A comprehensive national PNT system refers to the PNT service infrastructure utilizing a variety of PNT information sources based on different principles under the control of cloud platform (Yang 2016). In the future, pulsars may be the PNT

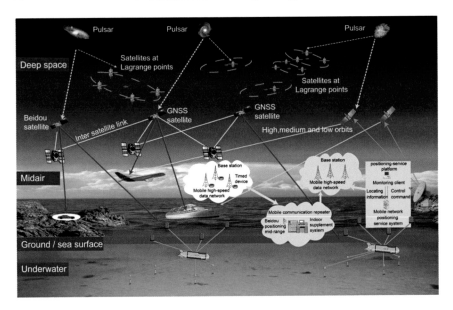

Fig. 4.4 Infrastructure framework of comprehensive national PNT system

information source for deep space detection; the satellite constellation at Lagrange points may provide radio PNT information with the same source and datum as the BeiDou system for the solar system and Earth-Moon system; the low Earth orbit (LEO) communication satellites would serve as the information source for low-orbit space; a PNT reference network similar to the geodetic network and satellite constellation can be built on the seafloor. Thus, a PNT infrastructure network with seamless coverage from deep space to deep sea can be completely established (Fig. 4.4).

The comprehensive PNT system will be the important direction of space–time datum construction for China and some developed countries. The key issue involved is to form a unified space–time datum using various information sources based on different physical principles. Firstly, a radio celestial reference frame will be established through observing stellar and occultation, and the relationship between the optical celestial and other reference framework must be established. Secondly, through the measuring and modeling of high-precision Earth orientation parameters (EOP), the relationship between the geodetic datum and deep space datum will be established. Thirdly, through the joint observations of land surface, sea surface and seafloor datum points, the geodetic and marine datum will be related. Thereby, the datum unification of deep space, geodetic, deep Earth, and deep sea will be accomplished.

4.6.1   Collaborative PNT Systems in High, Medium, and Low Orbits

As for deep-space PNT satellite constellations, the Lagrange points of the Sun-Earth and Earth-Moon system are the optimum relay stations for PNT services, and navigation satellite constellations (high-orbit constellations) can be deployed. Lagrange satellites may be equipped with upper and lower antennae directing to the deep space and Earth, respectively. The upper antenna can observe pulsars and broadcast ephemeris for deep space vessels to provide PNT services (Fig. 4.5). The lower antenna can receive the medium and low orbit signals of GNSS and provide PNT services for near space vessels. Signals from Lagrange constellation and GNSS can be combined and operated synchronously (Li and Fan 2016; Gao et al. 2016; Lu 2012).

The main challenges faced by Sun-Earth and Earth-Moon Lagrange navigation constellation are the stability and autonomous controllability of the constellation itself; the autonomous orbit and clock bias determination of the constellation; the compatibility and interoperability of deep-space and low altitude constellations.

The MEO and LEO communication and remote sensing satellite constellations are suitable navigation augmentation constellations. They can not only provide enhanced PNT services, but also determine the Earth's gravity field (using high-low tracking or low-low tracking technology) and research spatial atmosphere (using radio occultation). The working patterns of LEO navigation constellation can be divided into two types. The first is broadcasting navigation signals directly from LEO satellites with small on-board atomic clocks. However, it is relatively difficult to construct the constellation, apply for operating frequency and control the cost. The second is providing ranging and positioning information using time-scale data carried along with communication signals from LEO satellites. In this way, on-board atomic clock

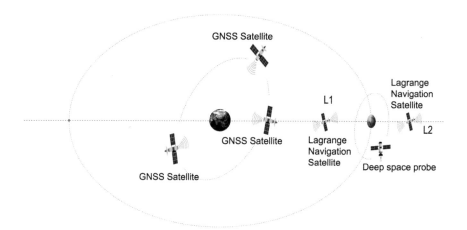

Fig. 4.5 Sketch of Lagrange navigation constellation

is unnecessary and the time information can be acquired by forwarding GNSS time. Whereas, it is incapable of providing independent, high-precision PNT services and can only be used to supplement and enhance medium- and high-orbit GNSS services.

As the supplement and enhancement of medium- and high-orbits GNSS constellations, low-cost LEO Earth observation and communication satellites would considerably increase the number of visible satellites to users, optimize the observation geometry and improve the precision of GNSS satellite orbits (Reid et al. 2016; Yang et al. 2020). In addition, the signal of LEO satellites would be relatively stronger and has better anti-interference ability (except deliberate interference). As a navigation augmentation constellation, low-orbit satellites could broadcast correction information, shorten initialization time for precise point positioning, and increase the calculation efficiency (Zhang and Ma 2019; Tian et al. 2019). Collaborative PNT service systems with high-, medium-, and low-orbit satellites will become a new focus of PNT services.

4.6.2 Pulsar Space–Time Datum

Pulsars are neutron stars that revolve at enormous speeds with extremely stable rotation frequency. Like the fixed star catalogue, the catalogue of pulsars constitutes a high-precision inertial reference frame. Pulsars generally send out periodic pulse signals, and the long-term stability of the rotation frequency of some millisecond pulsars is comparable to the most stable caesium atomic clocks. Therefore, pulsars can serve as natural clock and beacon in the universe that provide stable and reliable space–time reference for the navigation in deep space.

Evenly distributed pulsars can constitute a constellation similar to that of a navigation satellite system, and provide high-precision and high-security autonomous navigation for deep space craft. They can also serve as precise natural clock group and establish new time frequency datum with better stability. It can be further employed to identify atomic time shifts and unknown changes and to study tiny variations in Earth's rotation and other solar system ephemeris.

The difficulties of constructing the space–time datum of pulsars lie in generating high-precision catalogue to satisfy the deep space PNT service requirements and developing deep space signal receiving and processing sensors for pulsars with low power consumption.

4.6.3 Marine PNT System

The establishment of a marine datum requires the formation of a unified land-sea geodetic datum network combining the coastal and island continuously operating reference stations (CORS), sea level GNSS buoys, and seafloor datum points. The unification of land and marine space–time datum is realized through the joint

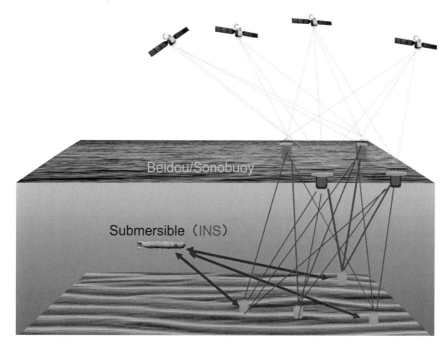

Fig. 4.6 Sketch of ocean datum observation network

measurement of geodetic, sea surface and seafloor control points and fusion of multi-source information (Fig. 4.6).

A seafloor geodetic datum network is generally established by reasonably deploying the sonar beacon network on the seafloor. In order to prolong the operational life of seafloor beacons, they can be designed to be purely passive or to operate in the 'wake-up' mode, which is to say they are normally in a dormant state but can be remotely activated at any time to provide navigation and positioning services.

There are a number of key problems involved. With regard to marine geophysics, marine geodynamic effects, subsea plate movements, subsea dynamic environmental effects, and optimal seafloor datum network geometry must be considered. With regard to seafloor beacon development, the beacons and shelters should be able to resist pressure, corrosion, and turbulence. As for PNT information transmission, sea surface observation geometry, seafloor datum network geometry, and observation equation establishing and observation fusion based on different physical principles are involved. With regard to marine geophysical inversion, fundamental theories and practical and scientific approaches must be developed for maintaining the coordinate accuracy of seafloor datum networks and datum points.

4.6.4 Miniaturized and Chip-Size PNT

Comprehensive PNT information sources are crucial for intelligent Earth, intelligent ocean and intelligent life. With the coexistence of various information sources, it's inevitable to realize comprehensive PNT sensor integration and information fusion. Otherwise, large, heavy and power-consuming user terminals will be incompatible with the principles of 'intelligent' cities and living. PNT application development must therefore aim towards developing chip-sized, modularized, low-power, integrated, miniaturized PNT terminals (Yang and Li 2017).

Micro-PNT offers not only micro-size (miniaturized) PNT equipment, but also "high precision + stability + reliability" in operation. Miniaturization demonstrates the optimized design principles, delicate manufacture techniques, and the deep integration technology for multiple micro-components. In addition, miniaturization requires the autonomous calibration of each measurement component, including both the active and passive calibration capacities, and also demands that the output information from various components in the PNT system be fused adaptively (Yang and Li 2017).

With sensor integration and miniaturization technology, the objective of Micro-PNT is to develop highly available, anti-interference, portable, stable, and low-power integrated sensors. The challenge to achieving goals is to develop inertial navigation components with ultra-stability and small accumulated error (e.g., quantum inertial navigation assemblies) based on super-stable chip-size atomic clocks.

4.6.5 Theory and Technology of Quantum Satellite Positioning

Pulse signals with entanglement and compression properties can be applied in quantum positioning. The pulse width and power (the number of photons contained in each pulse) are key factors affecting the positioning accuracy. The more photons contained in each pulse (equivalent to redundant observation), the higher measurement accuracy of pulse delay may be achieved (Bahder 2004).

As satellites with quantum entanglement capacity can only send out entangled photon signals to assigned position, they cannot provide PNT services in broadcasting mode like BeiDou or GPS. Therefore, it is not feasible to transmit quantum entangled photons from satellites, but from user terminals. The LEO satellites (with known orbits) receive and reflect the pulse signals transmitted by user multi-antenna transmitters, then the users calculate the distance between the transmitter and the satellites with the entangled to-and-fro pulse signals between them and thus calculate the user's position (Fig. 4.7). Possessing both high measurement accuracy and strong anti-interference and anti-deception abilities, quantum positioning technology can provide high-precision and high-reliability navigation and positioning services for core carriers in confrontation. At present, the quantum ranging accuracy using

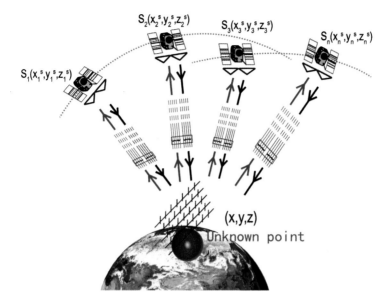

Fig. 4.7 Sketch of positioning principle of quantum satellites

single photon entanglement is better than 1 mm, and the double-photon measurement accuracy is expected to reach 1 µm, which could provide sufficiently high-precision data support for real-time monitoring of plate motion (Pan et al. 1998, 2003; Liao et al. 2017; Giovannetti et al. 2001, 2002; Jozsa et al. 2000; Peters et al. 2001).

The key fundamental theories of quantum satellite positioning include quantum entanglement theory and devices for the user terminal, antenna design of the user terminal, the pointing accuracy of laser transmission and its influence, the model of energy consumption and the design of quantum positioning constellation.

4.6.6 Datum Measurement Technology for Optical Clocks and Elevations

The traditional leveling and gravity measuring methods for height datum construction are inefficient, labor intensive and error cumulative, which cannot be employed in mountains or sea areas. Thus, the elevation datum obtained is regional and difficult to be unified globally.

According to the Einstein's theory of general relativity, clocks are affected by the gravity fields. With a common time scale, the clock operates at different speeds under different gravitation, which means that the frequency shift of atomic clocks is closely related to the gravitational potential difference. Thus, the frequency shift between local and remote (optical) clocks can be determined according to the gravitational

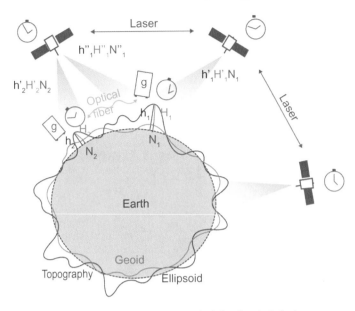

Fig. 4.8 Sketch of the height datum measurement principle of optical clock

potential and height differences between the two points and, in turn, the gravitational potential difference can be determined with the clock frequency shift (Shen et al. 2016, 2017, 2019a, b; Mehlstäubler et al. 2018; Chen et al. 2016; Freier et al. 2016). The spatial gravitational potential and height variation can be calculated through precise monitoring of the time–frequency transitions of high-precision atomic clocks at different locations, thereby accomplishing the direct measurement of gravitational potential differences and achieving the unification of global height datum (Fig. 4.8).

The height measurement using optical clocks is a revolutionary technology for the spatial datum establishment. The key factors involved include the high-precision atomic clock technology and the long-distance frequency comparison technology. At present, the precision of optical atomic clocks is up to 6.6×10^{-19}/h, which can sense elevation variations of about 7 mm. If the frequency comparison accuracy reaches 1×10^{-18}, the accuracy of elevation measurement can be within 1 cm. A small optical clock will be equipped in the China's space station to explore the feasibility of the large-scale unified height datum (Fig. 4.9).

The fundamental theories of the height datum establishment with optical clock include the precise theory relations and models between gravitational potential difference and clock frequency, high-precision time transfer strategies and algorithms of optical clocks at different sites (the required transfer error is much smaller than the error of the optical clock itself), the frequency error influence models of various optical clocks on gravitational potential difference, and the influence mechanism and quantitative evaluation of other disturbances on the optical clocks frequency.

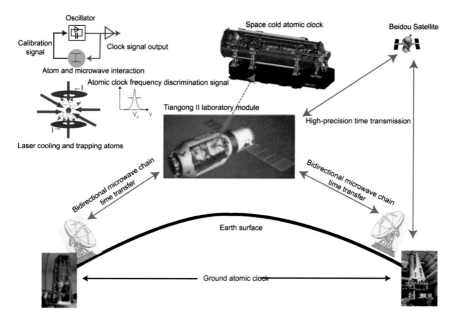

Fig. 4.9 Measurement of photonic height datum based on space stations

4.6.7 Technology and Fundamental Theory of Resilient PNT

Based on comprehensive PNT information, resilient PNT takes the optimized and integrated multi-source PNT sensors as the platform, adopts resilient adjustment of functional models and resilient optimization of stochastic models, and realizes the integration and generation of available PNT information across a range of complex settings. Thus, the high availability, continuity, and reliability of PNT information are ensured, and so is the safe and reliable application of PNT in critical infrastructure (Yang 2019).

The PNT services involved in national critical infrastructure operation and personal security must be safe, reliable and continuous. However, any single PNT information source has the risk of unavailability, which may affect the safety and integrity of PNT service. Thus, resilient PNT is proposed to realize the "resilient" selection of PNT information through polymerizing redundant information, and improve the reliability, security and robustness of PNT services for various carriers.

Resilient PNT is a revolutionary technology for PNT integration and application. The fundamental theories involved are resilient integration of PNT sensors, resilient functional modeling, resilient data fusion theories and methods (Yang 2019), and resilient application of PNT infrastructure. The key technologies include the adaptive identification of observation model and the environment, resilient observation modeling adaptable for different environments, theoretical models of resilient data fusion, etc.

4.7 Big Data and Artificial Intelligence

The digital revolution, represented by big data and artificial intelligence, is profoundly altering our life and the way of thinking, and will inevitably change scientific research paradigms. Over the past two or three centuries, a series of industrial revolutions has significantly speeded up the process of civilization and facilitated the development of science and technology. It is believed that the main driving force for the current and near future scientific and technological breakthroughs and the development of a sustainable society is the digital revolution. As a matter of fact, scientific research has entered the fourth industrial revolution marked by the new paradigm of data-intensive knowledge discovery. The principal digital technologies including artificial intelligence, the internet of things, 3D printing, virtual reality, machine learning, blockchain, robotics, and quantum data processing will create unprecedented opportunities to solve many major and complex problems faced by modern science and society.

Multiple examples demonstrate the two unique advantages of big data analysis and artificial intelligence. One advantage is the increasing volume of available data, while data collection, arrangement, cleaning, analysis, and other preparation processes can be simplified and optimized so that data acquisition is no longer time and labour intensive. The other advantage is the discovery of previously unrecognized processes and scientific patterns or rules by using data and knowledge mining, resulting in a rapid evolution from the scientific and technological paradigm of 'searching answers for known scientific questions' to 'searching for unknown answers for unknown questions' (Cheng et al. 2020).

Machine learning, machine reading, complex networks, and knowledge graphs can collate and integrate relevant data from multiple sources in diverse disciplines to achieve effective digital association and visual presentation of complex systems and interactions between numerous spheres, scales, media, and processes in the Earth. Every significant extreme event occurring within Earth's complex system can be analyzed deeply, associated remotely, simulated quantitatively, and predicted statistically using knowledge mining based on deep machine learning, artificial intelligence, and complex systems analysis. Big data and artificial intelligence will be utilized in the geosciences to facilitation of vital data-driven geoscientific discovery.

4.7.1 Data Integration, Data Assimilation, and Knowledge Sharing

To carry out research on the Earth habitability under the vision of "three deep and one surface" systems (deep Earth, deep ocean, deep space, and surface systems), it is necessary to collect geoscientific data from diverse disciplines, on multiple spatiotemporal scales, across enormous ranges, and in vast volume. An integrated Earth monitoring system must be constructed with strong capacity for observation and data processing that can link diverse elements and processes involved in the land-, space-, and ocean-based monitoring systems. Data assimilation technology should be developed to form reusable datasets that are continuous in time and space, with consistent internal association and the capacity for describing co-evolution of sea-land-atmosphere systems. Just to name a few examples, construction of databases including sediment geochemical database that can spatially or temporally link sedimentological, biological, and environmental information is necessary for holistic reconstruction of the evolution of surface processes in the Earth system; a fundamental dataset of global environmental changes is needed for comprehensive study of the environment and associated hazards; deep space exploration needs panoramic detection around the entire Sun using measurements from the solar-Earth sphere networks; a ground platforms-based "digital space brain" system should be developed; and for deep Earth research, one of the focuses should be given on establishment of globally unified time scale by combining global geologic time variation data and other relevant criteria.

The explosive growth in geoscientific data resources has resulted in a variety of large-scale and distributed professional databases as well as the rapid development of data analysis and computational infrastructures. For instance, engineering databases can provide structured spatial data in a machine-readable form, whereas scientific data that are collected, preserved, maintained, and partially published by scientists are dispersed in various media (e.g., journal papers, books, research reports, and web platforms) as text, figures, photographs, tables, audio, and video. Most of these unstructured data cannot be easily read, queried, or integrated by machine. These types of scientific data are complementary and must be integrated to describe the entire Earth. Technically, the principal difficulties in geoscience data utilization are data assimilation and database interoperability and compatibility. With respect to data that is still not shared, await utilization after release, or is not automatically readable by machine, global cooperation is required to promote data standardization, unification, and integration. A fundamental step in promoting data sharing is to ensure that data meets FAIR (Findable, Accessible, Interoperable, and Reusable) specifications and standards set down by the international scientific community.

4.7.2 Geoscientific Knowledge Graphs and Knowledge Engines

Obtaining information from data by scientific, rational, and effective methods is the principal concern of geoscientists to aid in understanding and interpreting geoscientific issues. Therefore, a considerable criterion for quantification of the results of big data processing is how much useful information and knowledge can be extracted from the big data.

The Earth is a complex system. The spheres of the Earth are in a continual process of dynamic interaction and mutual influence. Fundamental objectives of

Fig. 4.10 Integration of geosciences, computer technologies, engineering and mathematics (Cheng et al. 2020)

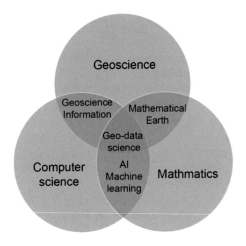

current geoscience are to study the operating mechanisms, relationships, and controlling and adjustment factors of the co-evolution of the spheres. Multi-disciplinary efforts are required to address this fundamental and universal geoscientific question. Driven by this and other fundamental questions, the geosciences are undergoing a transition from seeking answers to specific questions in individual disciplines to addressing universal and inclusive questions in transdisciplines. A deep integration of modern disciplines (e.g., data science and mathematical geosciences) and traditional geoscience disciplines is required to break down current interdisciplinary barriers and to construct a new theoretical knowledge system to connect all geoscience disciplines. New data analytical technologies are essential to deal with and analyze big data to obtain new knowledge. Platforms connecting knowledge systems from multiple disciplines using digital pattern detection will achieve full integration of disciplines and provide the data needed for multi-dimensional geoscientific practice (Fig. 4.10).

Methods and technologies for swift and efficient big data processing are required for the integration of massive volumes of data from various disciplines and to deal with multi-source, heterogeneous, and rapidly accumulating data. Advanced information technologies (e.g., cloud and parallel computing, supercomputing, and complex networks) can provide crucial support. Knowledge resources can be described visually using knowledge graphs, and knowledge engines can provide intelligent knowledge management systems. Both knowledge graphs and knowledge engines can be applied in mining, analyzing, constructing, and displaying knowledge associations to achieve multi-discipline integration. Modern AI technology and advanced semantic knowledge engines can be jointly utilized to develop knowledge graphs to aid and accomplish the automat integration of human thinking and machine learning, and even extend to the processes of integrating machine learning, information management, and infrastructure networking. Knowledge engines can be constructed using specific or natural programming languages and are editable, amendable, expandable, and reusable. Knowledge graph and engine models can therefore be continuously improved and optimized by positive feedback processes.

Integration of big data and computing technologies will promote abductive reasoning driven by both data and physical models, thereby accomplishing organic integration of essential computing resources and human intelligence to deal with the challenges presented by complex geoscientific problems.

4.7.3 Deep Machine Learning and Complex Artificial Intelligence

Artificial intelligence is the simulation of human intelligence and behavior in computer software. The concept of machine learning was originally presented by the American computer and artificial intelligence expert Arthur Samuel and ever since it has been an important field in the research and development of artificial intelligence. The general objective is to extract characteristics and information automatically from raw data using computers. Artificial intelligence and machine learning are rooted in cross-fertilization between applied mathematics, applied statistics, and computing technology and have developed rapidly alongside big data technology and the recent exponential growth in computer processing power. In the past five years, the AlphaGo AI program developed by Google DeepMind has defeated the world's best human chess player, and artificial intelligence has attracted widespread attention and general approval in society. In 2019, "quantum supremacy" was successfully demonstrated by Google—a major milestone in computing which heralds an entirely new development space for supercomputers. The computing power of supercomputers and big data will boost technical development of complex artificial intelligence, deep machine learning, and machine reading.

Big data analysis and supercomputing can simulate, calculate, and predict Earth development, evolution, and sphere interactions. The internal mechanisms of, and connections among, important geological events can be revealed, and the future development trend of the Earth's systemic processes can be analyzed and predicted. In the deep sea, breakthrough research has already been carried out on prediction and forecast based on big data and artificial intelligence technologies. Technological support from both artificial intelligence and machine learning are necessary for research on the evolutionary laws of habitable environments on terrestrial planets, the development of intelligent identification technologies for natural catastrophes driven by analysis of disaster mechanisms and data synthesis, construction of a digital model of the ionosphere system, and the construction of digital terrestrial space models with high spatiotemporal precision.

The capacities of artificial intelligence and machine learning programs depend on mathematical models, algorithms, and computer processing power. At present, calculation methods commonly used in machine learning include logistic regression analysis, artificial neural networks, naïve Bayes classifiers, random forest algorithms, and adaptive enhancement, all of which are at the forefront of research and development (R&D) and application development in mathematical geosciences. For instance, neural network models are becoming a mainstream element in machine learning utilized in quantitative prediction and assessment of mineral resources and energy by academia, governments, and industry. However, complex geoscientific problems involve nonintegrable or nondifferentiable functions (*e.g.* fractal curves or curved surfaces) that can't be approximated infinitely by current neural networks. Approximation theorem proves that only the Lebesgue integral function can be approximated in neural networks, which ensures infinite approximation of geological problems as long as they can be represented by smooth, continuous, integrable functions. For dealing with geological problem represented by complex functions, artificial neural networks with greater width and depth must be utilized which usually generate a wider range of model parameters requiring more data and computing power. The results are also unstable or non-convergent, and their repeatability and precision are also significantly reduced. These represent significant constraints on the use of artificial intelligence in research on Earth complexity. The currently fragmentary observational database is another issue for the application of artificial intelligence in the geosciences. Innovation in theories and methods is still imperative in the design and modelling of neural networks targeted on extreme Earth events caused by abrupt system changes. This type of artificial intelligence can be called 'complex artificial intelligence' or 'local precision artificial intelligence'.

4.7.4 Data-Driven Geoscientific Research Paradigms Transformation

The digital revolution will promote two major transitions in geoscientific research: the transition from expert learning to machine learning and artificial intelligence, and scientific paradigm transition from problem driven to data driven.

Human beings obtain new knowledge and experience by re-learning and re-practicing previous experience and knowledge. Machine learning and artificial intelligence can effectively promote positive feedback processes to improve the performance of the learning system itself. Primary objectives of machine learning and artificial intelligence are extracting the patterns hidden in data, and uncovering new knowledge by data analysis. New functions and advanced behavior of robots can be achieved by integrating advanced technologies like computer vision and natural language in machine learning and artificial intelligence. These types of technologies have been rapidly applied in many fields of natural and social sciences, and engineering, and profoundly affecting industries. Machine learning and artificial intelligence technologies have also been utilized successfully by geoscientists in quantitative analysis, simulation, prediction and evaluation of extreme phenomena such as severe weather events, volcanic activity, earthquakes, and mineral and energy resources. This indicates that machine learning and artificial intelligence have a huge potential in the detection, simulation, and prediction of extreme events based on geoscientific big data. The development of big data and artificial intelligence must

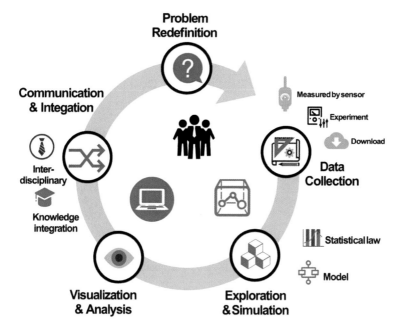

Fig. 4.11 Major components of integrated data, models, algorithms, data mining processes, computer networks and personnel to support data-driven knowledge discovery (Cheng et al. 2020)

be leveraged to promote the integration and cooperation of machine learning, human intelligence, and artificial intelligence (Fig. 4.11).

From the perspectives of Earth system science and habitable Earth, the main issues confronting geoscientists can be classified into two groups. One group consists of fundamental scientific issues relating to the interaction and co-evolution of the multiple spheres in the Earth system. The other group consists of application issues related to the mechanisms, prediction and spatial–temporal distribution of natural resources (*e.g.* mineral resources, energy, and water) and of geological hazards (*e.g.* earthquakes, volcanoes, landslides, flooding, tsunamis). Based on geoscientific big data, super-computing capacity, and modern artificial intelligence technologies, researchers can correlate and analyze lithosphere, biosphere, hydrosphere, and atmosphere under a unified geological age "deep time scale" including exploring the variations, internal correlations, and co-evolution of every relevant sphere in the spatiotemporal scale of the Earth. For example, the use of time series analysis or complex network techniques to analyze global data sets can quantitatively describe the changing trends of the Earth's biodiversity, the chemical index of the atmosphere, the chemical composition of seawater, and the changing laws of some mineral phases. Reconstruction of Earth's evolution thereby reveals the complex correlations between important and extreme geological events (e.g., supercontinent formation, massive biologic extinctions, BIF iron ore body formation, the Great Oxidation Event, snowball Earth event, and large-scale mineralization). A number of crucial discoveries may follow, including identification of unrecognized long distance interdependent relationships (symmetric or asymmetric) between vital events, and understanding of mutual cascade circulation and gradually upgrading effects. Data-driven study can address these unanswered questions, providing breakthroughs in the research paradigm of 'finding unknown answers to known questions'.

References

Bahder TB (2004) Quantum positioning system. In: Proceedings of the 36th annual precise time and time interval systems and applications meeting, Washington, D.C., December 2004, pp 53–76

Chen LL, Luo Q, Deng XB et al (2016) Precision gravity measurements with cold atom interferometer (in Chinese). Sci Sin-Phys Mech Astron 46:073003. https://doi.org/10.1360/SSPMA2016-00156

Cheng QM, Oberhänsli R, Zhao M (2020) A new international initiative for facilitating data-driven Earth science transformation. Geol Soc Lond 499(1):225–240

Freier C, Hauth M, Schkolnik V et al (2016) Mobile quantum gravity sensor with unprecedented stability. J Phys: Conf Ser 723:012050

Gao YT, Xu B, Zhou JH et al (2016) Lagrangian navigation constellation seamlessly covering lunar space and its construction method (in Chinese). PRC Patent, CN105486314A, 2016-04-13

Giovannetti V, Lloyd S, Maccone L (2001) Quantum-enhanced positioning and clock synchronization. Nature 412(6845):417–419

Giovannetti V, Lloyd S, Maccone L (2002) Positioning and clock synchronization through entanglement. Phys Rev A 65(2):022309

Jozsa R, Abrams DS, Dowling JP et al (2000) Quantum clock synchronization based on shared prior entanglement. Phys Rev Lett 85(9):2010–2013

Li M, Fan JJ (2016) Analysis of the lunar navigation constellation scheme (in Chinese). Geomatics Sci Eng 036(005):20–23

Liao SK, Cai WQ, Liu WY et al (2017) Satellite-to-ground quantum key distribution. Nature 549:43–47

Lu Y (2012) Design and analysis of lunar communication relay and navigation constellation (in Chinese). Univ Electron Sci Technol China, Chendu, p 126

Meadows VS, Barnes RK (2018) Factors affecting exoplanet habitability. In: Deeg HJ, Belmonte JA (eds) Handbook of exoplanets. Springer, Cham, Switzerland

Mehlstäubler TE, Grosche G, Lisdat C et al (2018) Atomic clocks for geodesy. Rep Prog Phys 81(6):064401

National Academies of Sciences, Engineering, and Medicine (2018) Exoplanet science strategy. The National Academies Press, Washington, DC. https://doi.org/10.17226/25187

Pan JW, Bouwmeester D, Weinfurter H et al (1998) Experimental entanglement swapping: entangling photons that never interacted. Phys Rev Lett 80(18):3891–3894

Pan JW, Gasparoni S, Ursin R et al (2003) Experimental entanglement purification of arbitrary unknown states. Nature 423:417–422

Peters A, Chung KY, Chu S (2001) High-precision gravity measurements using atom interferometry. Metrologia 38(1):25–61

Pierrehumbert RT (2010) Principles of planetary climate. Cambridge University Press, Cambridge, p 674

Reid TG, Neish AM, Walter TF et al (2016) Leveraging commercial broadband LEO constellation for navigation. In: Proceeding of the 29th international technical meeting of the satellite division of the institute of navigation (ION GNSS + 2016). ION, Porland, Oregon, pp 2300–2314

Shen ZY, Shen WB, Zhang SX (2016) Formulation of geopotential difference determination using optical-atomic clocks onboard, satellites and on ground based on Doppler cancellation system. Geophys J Int 206(2):1162–1168

Shen ZY, Shen WB, Zhang SX (2017) Determination of gravitational potential at ground using optical-atomic clocks on board satellites and on ground stations and relevant simulation experiments. Surv Geophys 38(4):757–780

Shen WB, Sun X, Cai C et al (2019a) Geopotential determination based on a direct clock comparison using two-way satellite time and frequency transfer. Terres Atmos Ocean Sci 30:21–31

Shen ZY, Shen WB, Peng Z et al (2019b) Formulation of determining the gravity potential difference using ultra-high precise clocks via optical fiber frequency transfer technique. J Earth Sci 30(2):422–428

Tian Y, Zhang L, Bian L (2019) Design of LEO satellites augmented constellation for navigation (in Chinese). Chin Space Sci Technol 39:55–61

Wu Z (2016) Research on control of ventilated supercavitating vehicles based on water tunnel experiment (in Chinese). Haerbin, Master Thesis, Harbin Engineering University, 73

Yang YX (2016) Concepts of comprehensive PNT and related key technologies (in Chinese). Acta Geodaetica et Cartographica Sinica 45(5):505–510

Yang YX (2019) Resilient PNT concept frame. J Geodesy Geoinf Sci 2(3):1–7

Yang YX, Li XY (2017) Micro-PNT and comprehensive PNT (in Chinese). Acta Geodaetica et Cartographica Sinica 46(10):1249–1254

Yang YF, Yang YX, Xu JY et al (2020) Navigation satellites orbit determination with the enhancement of low Earth orbit satellites (in Chinese). Geomatics Inf Sci Wuhan Univ 45:46–52

Zhang X (2020) Atmospheric regimes and trends on exoplanets and brown dwarfs. Res Astron Astrophys 7:1–92

Zhang XH, Ma FJ (2019) Review of the development of LEO navigation-augmented GNSS (in Chinese). Acta Geodaeticaet Cartographica Sinica 48(9):1073G1087

Chapter 5
Realization: Intersectionality, Integration, Collaboration, and Cooperation

5.1 Platform Construction and Data Sharing

There are various ways to acquire data in Earth science research. Observation and timely sharing of datasets are indispensable approaches for deepening the quantitative research of the Earth system. Shortcomings need to be identified with respect to existing observation systems for each sphere in order to construct an integrated observation system for sea-land–atmosphere interaction that is interrelated, internally coordinated, connected into a network, and unified and standardized. This will promote the development and application of a new-generation of Earth observation technologies (e.g., internet of things (IoT), sensors, and telemetry observation), and achieve a continuous all-weather multi-sphere and multi-element observation system for the Earth system.

In terms of information sharing, long-term field observations and experimental survey data must be collected to act as the foundation. Data from social statistics, remote sensing and inversion, aerial observations, surface surveys, model simulations, and other data resources should be harvested and integrated to form standardized model-driven datasets, parameter datasets, and verification datasets. Sharing mechanisms must be established to generate a set of continuous, stable, and standardized scientific observation data of the Earth system and relevant economic and social data as well. A sharing platform for scientific and technological data and information systems must be constructed. A thematic information sharing platform for comprehensive scientific research on typical regional field observations, comprehensive surveys, laboratory experiments, process simulations, and scenario and decision analysis should be formed, and a resource sharing platform should be established for scientific data acquisition, comprehensive analysis, knowledge innovation, and science popularization.

© Science Press 2022, corrected publication 2022
The Research Group on Development Strategy of Earth Science in China,
Past, Present and Future of a Habitable Earth, SpringerBriefs in Earth System Sciences,
https://doi.org/10.1007/978-981-19-2783-6_5

5.2 Interdisciplinary and Collaborative Research

Interdisciplinary integration should be strengthened. The study of deep-Earth processes and a habitable Earth involves two core disciplines, solid and surface Earth sciences. Breakthroughs will only be made by taking the broad perspective of "deep Earth, deep space, deep sea, and the Earth system". The Earth is a vast and complex system. Along with continual differentiation of disciplines, reinforcing interdisciplinary and encouraging collaborative research is the fundamental way to cope with the complexity of the Earth system. Relying on combinations of research methods (e.g., natural rock sample testing and analysis, HPHT experiments, geophysical exploration, and multi-scale computational simulation), the physical, chemical, and biological processes and mechanisms of their interactions in major geological events will be revealed by constructing databases of various records of related geological events from a global perspective. This will provide a scientific basis for understanding the formation and evolution of the Earth's habitability. The disciplines involved in the study of Earth's habitability include not only those related sub-disciplines of Earth sciences, but also other disciplines such as life sciences, technical sciences, and social sciences. The human-Earth system is a coupled system formed by the interaction between human beings and natural systems, which can only be studied using an interdisciplinary approach involving natural sciences, social sciences, humanities, and information technology.

Leveraging the advantages of the Division of Interdiscipline of the National Natural Science Foundation of China, multi-disciplinary and multi-field research teams can be organized to identify new research paradigms and perform comprehensive research on "deep Earth, deep space, deep sea, and the Earth system" from the global perspective of understanding the past, present, and future of the habitable Earth, and to improve the collaborative research capacity in science and technology. Based on this approach, the systems of national funding and relevant project funding should be further developed and improved to offer a strong combination of competitive selection and stable support. New models can be added to existing funding schemes, and increasing the investment of the National Natural Science Foundation of China should place a strong emphasis on basic scientific research. From the aspect of the national talent pool, it is necessary to break the boundaries of disciplines, strengthen support for interdisciplinary research, and cultivate interdisciplinary talents. On the other hand, it is also necessary to actively recruit or cultivate top scientists and technicians in specific fields who possess a global perspective.

5.3 International Collaboration and Exchanges

At present, there is still a large gap between scientific and technological levels in China and those in Europe and the United States. This reality will be changed in the short term. International scientific and technological collaboration and exchange must be encouraged, particularly between China and the United States.

A key issue to be urgently resolved in the Chinese scientific and technological community is to bring innovation and fresh thinking to international collaboration and exchange. We must put forward the strategic thinking of scientific and technological transformation in a spirit of cooperation and learning, shifting from 'imitation and tracking' to 'exploration and innovation'. This requires that Chinese scientists not only adopt an international perspective, but also understand the strategic needs of the country. At present, the development of science and technology is increasingly dependent on large-scale scientific projects and programs jointly organized and implemented by multiple organizations, departments, and even countries or regions. The inherent properties and characteristics of Earth science make international cooperation essential for the study of major scientific questions. Participation in, and then initiating and organizing major international scientific research projects will promote China to enter the forefront of global science, enhance the research level of Chinese Earth science, and improve the innovation capacities of the Earth science community in China. It will also ensure the contribution of Chinese wisdom to the development of international Earth sciences.

In recent years, international scientific and technological cooperation in Earth sciences, particularly the development of a series of major international initiatives, has achieved significant outcomes in talent exchange and training, hardware and infrastructure sharing, and tackling scientific problems. However, the overall environment for international cooperation in our country still requires further improvement, and cooperation experience and capabilities also urgently needs to be improved. The in-depth development of China's international scientific and technological cooperation is facing a number of serious challenges:

(1) **Administrative management**: The successful implementation of large-scale international scientific projects depends on the overall coordination and cooperation between individuals, institutions, departments and even countries. However, there are often clear regional and departmental boundaries in China, which makes it difficult to achieve cross-departmental joint organization and funding. Organization, management, and service models need to be improved and optimized. There are still significant shortcomings in the participation and integration of Chinese scientists into the international academic community.

(2) **The scientific community**: The consciousness, ability, and level of independent design and organization of international large-scale scientific projects are insufficient, and the ability to formulate rules in international cooperative research and sharing of results is lacking.

(3) **Subsidiary and evaluation system**: Long-term, systematic and comprehensive observation and research are required to solve major scientific questions in the field of Earth sciences. Most of the existing international cooperation projects in China are short-term and cannot provide sustained support. In addition, the evaluation system has yet to be fully aligned with international standards.

By focusing on system and method innovation, and striving to create a good atmosphere of openness and cooperation, these above-mentioned problems can be solved. It will surely promote China's active integration into the global science and technology innovation network, with a range of specific bilateral and multilateral cooperation efforts sponsored and performed systematically in order to lead the way toward constructing a global scientific and technological innovation community of mutual benefit and win–win cooperation.

Correction to: Past, Present and Future of a Habitable Earth

Correction to:
The Research Group on Development Strategy of Earth
Science in China, *Past, Present and Future of a Habitable*
Earth, **SpringerBriefs in Earth System Sciences,**
https://doi.org/10.1007/978-981-19-2783-6

In the original version of the book, the following belated corrections have been incorporated: The author name "The Research Group on Development Strategy of Earth Science" has been changed to "The Research Group on Development Strategy of Earth Science in China" in the Frontmatter, the Backmatter and in all chapters.

The updated original version of the book can be found at https://doi.org/10.1007/978-981-19-2783-6

© Science Press 2022
The Research Group on Development Strategy of Earth Science in China,
Past, Present and Future of a Habitable Earth, SpringerBriefs in Earth System Sciences,
https://doi.org/10.1007/978-981-19-2783-6_6

Postscript

The study of Earth habitability is a type of fundamental research, characterized by high investment, slow returns, and unquantifiable results. However, this research possibly defines the ability of a country to produce original and innovative work in applied Earth sciences. A nation needs to provide long-term, stable support at a strategic level. Meanwhile, researchers should also aim towards increasing their self-confidence, targeting cutting-edge topics, original theories, and revolutionary technology, achieving breakthroughs in the study of the Earth habitability and providing support in leading the fourth industrial revolution.

When did the Earth evolve into a habitable planet? What is the future direction of evolution of our planet? These are the core questions in geosciences and related fields in the twenty-first century. To understand the complex Earth system, we first need to develop innovative research paradigms. The first step is thinking outside of the box, meaning looking at the Earth system not just from the Earth's perspective, but also from inner Earth and outer space. Next, we take the time parameter into consideration, which leads us to unveil the past, focus on the present, and discover the future of the habitable Earth. The goal of research on habitable Earth is to serve biological sustainability. In recent years, bioscience has become the research area of greatest concern to the government and science community. For example, in 2005, *Science* selected 125 of the most important current scientific questions to commemorate the 125th anniversary of its publication, 46% of which were related to life sciences. In fact, biological health requires a habitable Earth, similar to 'the essence and the symptom' of Chinese traditional philosophy. The 'essence' is Geo-health, and the 'symptom' is traditional environmental protection, disaster prevention, medicine, etc. The fundamental problem faced by the mankind is to understand the coevolution and development of the Earth and life.

Deep space exploration has gradually enabled mankind to appreciate the sheer scale of the universe, and to realize that the Earth is "not special". This provides us new insights into the development of science and technology. In October 1957, the Soviet Union firstly launched the Sputnik 1 satellite, weighing 83.6 kg, into Earth orbit, which shocked the entire world. Subsequently the United States of America launched the lighter Explorer I satellite, weighing 14 kg, in February 1958. The Explorer I satellite carried cosmic-ray and microwave background detectors on board and consequently discovered the Earth's radiation belt (now known as the Van Allen belt).

© Science Press 2022
The Research Group on Development Strategy of Earth Science in China,
Past, Present and Future of a Habitable Earth, SpringerBriefs in Earth System Sciences,
https://doi.org/10.1007/978-981-19-2783-6

These discoveries have significant impacts on later studies regarding science and technology. Therefore, the space competition between the US and the Soviet Union shed light on the objective rules of harmonization between science and technology.

Deep space exploration is an important approach to understanding Earth habitability, offering a 'ventures lab' for explorers. In 1961, US President John F. Kennedy delivered an ambitious speech, proclaiming the national goal of "landing a man on the moon and returning him safely to the Earth". It was this innovative spirit and foresight of the American space team that delivered the magnificent achievement on July 20, 1969, the first Moon landing, which led to Neil Armstrong's famous words, **"That's one small step for man, one giant leap for mankind."**. To understand the past, present, and future of the habitable Earth, we need to generate scientific originality from multi-interdisciplinary studies. This is the most effective measure to make up for the past imbalance in the development of our science and technology. To start off, Earth science needs to extend the study of comparative planetary research. In the early days of the solar system, the conditions forming Venus, Earth, and Mars were remarkably similar in terms of solid composition, volatile matter, and external space environment, however the three planets have evolved very differently. In the 1960s, the United States, by sending exploration probes to Venus, discovered that the planet is in an out-of-control greenhouse state, which first aroused human concern about the warming of the Earth. Subsequent exploration has found that Mars, by contrast, is extremely cold and has only a thin atmosphere. These discoveries brought up some questions: Which way will the Earth go? How should we respond to changes in Earth habitability? To answer those questions, it requires us to 'explore the universe, understand, and make good use of the Earth', by implementing which we can achieve original theoretical breakthroughs, and at the same time serve the national development strategy.

What are the focus areas on universe exploration? What are the new ideas required to achieve breakthroughs in these areas? For instance, one of the fundamental goals of international Mars exploration is to find the evidence of life on Mars. What are the defining signs for the existence of life elsewhere in the universe? Are fresh water, oxygen, or air essential for life? It seems that we still cannot answer these basic questions. To answer those questions, innovative thinking may be a new way to determine the signs of life. Discussion of the question is often avoided. As a matter of fact, the complexity of life in the universe can be inferred from the diversity of life on Earth. For instance, most life on Earth requires air, sunlight, and water, yet submarine organisms in the deep oceans are usually anaerobic and photophobia. This stark contrast demonstrates that, on the Earth, the requirements of the existence of terrestrial and aquatic lives is significantly different. As we now understand that the Earth is not special, we may assume that there might be life forms elsewhere in the universe, generally similar to those on the Earth, but might be adapted to their specific environments. Another vital scientific tool for space exploration is to use comparative planetology to understand how the Earth's interior works. Because Mars currently has no internal dynamic processes, the planet's magnetic field has disappeared and most of the atmosphere has been lost. By studying the atmosphere of Mars, and samples returned by exploration probes, we can deepen our understanding of what happened when Mars' internal driving forces and self-regulatory functions stopped, which consequently will affect our understanding of how Earth habitability may evolve.

The Moon is a natural test site for space exploration and research on the influence of extraterrestrial bodies on Earth habitability. It is clearly essential that we study the co-evolution of the Earth-Moon system, therefore we should set out a scientific path towards building a lunar base. The former Soviet Union was the first nation to construct a space station, and China's own space station is now approaching completion. However, the Moon itself is, in fact, the ideal permanent 'space station' to support mankind's exploration of deep space. We therefore need to figure out life-suited conditions for astronauts on the Moon. We expect that, in the next 10–20 years, humans will better understand the Earth-Moon system and, evolution of Earth habitability, by first establishing unmanned research stations on the lunar surface.

Unlike the other planets in the habitable zone in the solar system, the dynamics of the Earth have produced plate subduction, volcanic activity, and frequent seismicity observed on the surface. It is the Earth's internal processes that provide the Earth with sufficient quantities of the renewable resources needed to support life. The reason why the Earth has evolved into its current state is directly related to the impact of plate subduction on the carbon cycle. In the early stages of the solar system, the CO_2 contents in the atmospheres of Earth, Mars, and Venus were roughly the same. Now, the Martian atmosphere is 95% CO_2 and the atmosphere of Venus is 97% CO_2, while Earth's atmosphere is only 0.04% CO_2. The main factor that contributes to this stark difference is that subduction of the Earth's oceanic plates transports CO_2 from the surface into the deep parts of the Earth. Fluids and volatile matter play important, but little understood, roles in this carbon transportation process which has not yet been solved by the theory of plate tectonics. Through the Earth's internal processes, the circulation of materials (including essential minerals, carbon, hydrogen, oxygen, and sulfur) among the deep Earth, the Earth's surface, and space has a decisive influence on Earth's habitability. The carbon cycle is a major controlling factor in the Earth's climate system. The hydrogen cycle is closely related to water and fossil fuel energy sources such as oil and gas. The oxygen cycle determines whether terrestrial lives flourish or die. The sulfur cycle is a controlling factor in the marine life system and is essential for producing a variety of mineral resources. Some studies have suggested that the interior of the Earth is a huge carbon reservoir. Perhaps the CO_2 emitted by humans is only a very short-term disturbance in the Earth's climate system. We cannot deny that human CO_2 emissions are having a significant impact on short-term weather fluctuations, but we must also note that the issues we are discussing are of a completely different level. In fact, the carbon content of the atmosphere and oceans accounts for only about 5% of recyclable carbon in the Earth system. We cannot ignore the 95% of the carbon stored in the Earth's interior when discussing habitability. Water is, of course, directly related to hydrogen, but all of the water contained in the vast Pacific, Atlantic, and Indian Oceans on the Earth's surface may still be insignificant compared to the water contained deep within the Earth.

In recent years, Chinese scientists have made exciting breakthroughs in identifying correlations between deep-Earth processes and the habitable Earth. For example, the team led by Prof. Mao Ho-Kwang found that the physical and chemical properties of substances with burial depths of less than 1800 km below the Earth's surface are entirely different to the properties of the same substances at greater burial depth. The inner zone (nearer the Earth's core) is controlled by thermal and mechanical processes different to the outer zone processes. These variations in two regions, within one system, can produce extreme potential energy difference, which could have provided the driving force for inducing major geological events throughout Earth's history. Taken hydrogen as an example, one of the main chemical carriers between the two zones is hydrogen, which is released by reactions involving hydrous minerals subducted into the inner zone. On the other hand, the oxygen produced by the reactions remains trapped in the inner zone. Superoxide is produced at the core-mantle boundary (CMB), and gradually accumulates at the bottom of the mantle, which may eventually induce chemogenic convection in the mantle, super mantle plumes, large igneous provinces (LIPs), extreme climate change, atmospheric oxygen fluctuations, biological extinctions, and other events. Familiar physiochemical laws show extraordinary changes in their properties under the temperature and pressure conditions of the lower mantle at depths of more than 1800 km. The study of these unknown physical states will foster innovative theories in geochemistry.

Facing the pressure of resources and energy shortages, and the current severe international situation, Earth scientists need to be forward-looking in setting out a new 'theory of the formation and enrichment of resources'. With an in-depth understanding of the 'foundation of territorial resources' gained from national projects such as 'Deep Earth Exploration', we urgently need research into the formation processes of newly discovered mineral types and proper resources extraction and utilization. For hydrocarbon exploration, we should look towards deep strata, deep water, and unconventional oil and gas, with a multi-disciplinary and integrated approach. Efforts should also be given to explore the new theory and technology of '*in-situ* conversion', and strive to lead the future fossil energy revolution. Thirty years ago, there was little optimism about the future of shale gas. However, technological progress facilitated and drove the 'shale gas revolution' at the beginning of the twenty-first century, which completely changed the structure of world energy resources. Shale oil and gas are now key unconventional resources. When and where will the next energy revolution emerge? Who will be the leaders? This is difficult to answer, yet it is certain that scientific and technological innovation is the key. We envisage that, in the future, the deep integration of Earth sciences with other basic sciences such as physics, chemistry, and mathematics will lead to innovation of new theories and technologies. For example, the accelerated maturation of abundant low-mature oil and gas resources-may be the key to the next energy revolution. Younger generations in particular need to be optimistic to the future. If they only believe it after seeing it, they will achieve no more than following in the footsteps of others. In the future, comprehensive research in multiple disciplines, such as Earth sciences, physics, materials science, chemistry, and chemical engineering will be essential to achieve breakthroughs in the theory and technology of *in-situ* shale oil conversion, which will be a major engine for the next energy revolution. The habitable Earth and the formation of minerals are also co-evolved. Two-thirds of the 5739 types of minerals on the Earth were formed after the emergence of life. The formation of new rock types is often closely related to biological activity. Studying the formation mechanisms of staple and rare mineral resources is not only a national strategic requirement, but also a new way to understand Earth habitability.

The oceans, which cover 70% of the Earth's surface, are a strategically vital area for the sustainable development of human society. Focusing on 'deep-sea entry, deep-sea detection, and deep-sea exploration', many nations are engaged in, or will embark on, major projects such as constructing deep-sea habitations (marine 'space stations'), oceangoing drilling vessels, E-class supercomputers, transparent oceans, and bathyscaphes. Scientists from different countries should think globally—from the perspective of the universe—and study fluid-solid interactions and submarine dynamic processes. As the Earth's 'blood', fluid/volatile matter not only connects the transfer of matter and energy with large-scale structural deformation, but also controls the transformation of the physical and chemical properties of the Earth system and the transformative characteristics of Earth habitability. Thus, the study of fluids/volatile matter is a vital starting point from which to observe and understand the interactions of Earth's multiple spheres and the planet's habitability.

One of the most important research achievements of oceanography in the past forty years is the realization that photosynthesis is not necessary for life on the sea floor. There are numerous species on the seafloor, at depths of several kilometers, which rely solely on chemical synthesis. In the past, it was generally accepted that everything depends on the sun for growth. However, even though the deep ocean is aphotic and hypoxic, organisms still flourish. Understanding where and how life on Earth emerged has always been a fundamental scientific problem. The ocean is a water-rock reactor. Serpentinite is one of the most abundant products of water-rock interaction on Earth. Serpentinization is the engine for heat generation and the structure of the Earth, and for the origin of life on Earth. The production of submarine hydrothermal fluids, especially low-temperature, alkaline, hydrothermal methane, provides us clues for describing the transformation of inorganic matter to organic matter. In the future, geoscientists, chemists, biologists, etc., will work together to study the generation of methane and various protein substances in mud volcanoes by serpentinization, which may help us understand the origins of life on Earth. More importantly, studying both the extinction and flourishing of living organisms from the perspective of the Earth's planetary system is essential for understanding the past, present, and future of the habitable Earth, as well as for planning how humans can eventually migrate safely to other planets.

The United Nations has proposed a series of sustainable development programs that focus on surface processes, in an attempt to understand the dynamic impact of the human-earth coupling system formed by the interaction of water, soil, air, life, and human activity. We want to build a charming China with ecological sustainability. To achieve this goal, we always need to consider China's economic development and urbanization in terms of the environmental capacity. The core scientific issue in this respect is related to Geo-health.

The history of scientific development illustrates that geoscience and geo-health are closely related. Geoscience researchers are all aware of the Danish scientist, Nicholas Steno's paper *Hard texture in natural solids* which proposed the basic principles of stratigraphy—the law of superposition of new strata on top of old strata, the law of continuous horizontal extension of stratigraphic deposits, and the law that strata occur horizontally in the absence of external change. These three laws are still valid until now. In fact, Steno was a Danish royal physician interested in Geology. The interest was driven by his curiosity and exploring spirit. Another great scientist that we know well is William Gilbert. He published his monograph, *De Magnete,* 400 years ago, firstly describing the Earth's geomagnetic field and challenging the traditional concept that the geomagnetic field is produced by God. Similar to Steno, Gilbert was also a royal physician interested in geomagnetism. Copernicus, the famous Polish Renaissance astronomer and mathematician, published his heliocentric theory of the solar system, *De Revolutionibus Orbium Coelestium,* in 1542, marking the beginning of the first scientific revolution and challenging the restraints of medieval theology and empirical philosophy. Copernicus used a modern natural science approach including observations, experiments, and mathematical analysis, and disproved the 1400-year-old geocentric model. Copernicus was also a physician, and was honored as a 'miracle-working' doctor. Galileo was the discoverer of gravity. He measured the acceleration of objects under Earth's gravity in 1590. Galileo also originally studied medicine before turning to mathematics. His hobby was to cure as many patients as possible, and he was regarded as an excellent physician. Xun Lu was also a medical student in his early career. He wrote the first treatise on geological structure in China, *A Brief Discussion on the Geology of China.* This apparent coincidence raises the question: "Why do doctors abandon their profession to "diagnose the Earth?" My view is that geo-health and Earth are bonded by coordinated development. If we want humanity to be healthy, we have to ensure that our planet is healthy. When facing new and unknown diseases, we gain a deeper understanding of the significance of studying and learning our nature. Meanwhile, we also realize that our existing knowledge system is inadequate to ensure the harmonious development of human race and the habitable Earth. The visit of unknown viruses makes us even more aware of the significance of geo-health for mankind, as well as the great responsibility that geoscientists take.

China's 40 years of integration into the world community shows that scientific and technological cooperation and exchanges are the key and core of international relations. As a big nation in the world, China needs to be open and tolerant with wisdom and self-confidence to deal with the complex international relationships. China should not dream of going back to its model of 40 years ago simply due to the temporarily unfriendly situations imposed by some countries. We should continue to improvise by trial-and error. Innovation in international cooperation and exchanges is a new challenge facing the science and technology communities. Chinese scientists need to propose strategic ideas and concepts for scientific and technological transformation in the context of international cooperation. We not only need global vision, but also need to understand the strategic needs of the nation. To understand the intrinsic attributes and characteristics of Earth sciences in order to tackle major scientific issues and to make efficient use of global innovative resources requires geoscientists to have a global vision and an ability to conduct extensive in-depth international cooperation and exchanges. At the high starting point of the community with a shared future for mankind, the international scientific and technological community needs joint efforts to make due contributions to the common challenges faced by mankind.

Academician of the Chinese Academy of Sciences: Rixiang Zhu

Beijing, 2021

Printed in the United States
by Baker & Taylor Publisher Services